Dancing on Water

Dancing on Water

Adventures with Dolphins, Whales and Interspecies Communication

Karin Kinsey

Dolphin Press 2005

DANCING ON WATER
Adventures with Dolphins, Whales and Interspecies Communication

A Dolphin Press Book
San Rafael, CA
www.dolphinpress.com

Layout and cover design by Karin Kinsey.

Publisher's Cataloging-in-Publication Data

Kinsey, Karin.
    Dancing on water : adventures with dolphins, whales and
interspecies communication / Karin Kinsey.
      p. : ill. ; cm.
    ISBN 0-9769282-0-5
1. Human-animal communication. 2. Human-animal relationships.
3. Spiritual life. 4. Dolphins—Anecdotes. 5. Whales—Anecdotes. I. Title.

BF1999 .K46 2005
133.9/3   2005904232

Printed and Bound in the United States
First Edition: November 2005

PHOTO SOURCES
John D. Kinsey: pages 5, 6
David Patterson: pages 19, 21
Mary-Anne Alderete: pages 20, 23, 51, 52
Karin Kinsey: pages 39, 44, illus. 76, 127, 129, 136
Photo Roatan: pages 57, 70, 72, 84, 110, 113, 117
Jon Cotton: pages 65, 107, 118
Fotosearch, LLC: pages 95, 98
Rolf Hicker/Photographer: pages 96, 101
Christina Craft/Photographer: page 97
Richard Iribarne: page 126

To Tela —
    with all my heart

ACKNOWLEDGMENTS

I give much thanks to my parents, my siblings, my friend and ally Jon Cotton, and all the dolphins and other animal companions who have been with me along the way. I would also like to thank those who have gone before and who through their love and dedication continue to devote time to the research, understanding and preservation of our cetacean friends. I thank the growing numbers of people who are involved in the conservation of our oceans, our reef systems, our water and our natural coastlines.

Finally, I extend my deepest gratitude to the many other people who have given assistance and support in the creation of this book, including Dawson Church, who with his generous heart and abundant expertise took me under his wing; Gary K. Smith, M.B.A. and Michael Mayer, Ph.D., who served as guides and mentors; Bodhi Setchko, my muse and inspiration; Marianne Betterly-Kohn for her insight and suggestions; and Marianna Cacciatore and Beverly Brunelle for their deep friendship and faith in this project. Lastly, I give special thanks to my editor, Larry Boggs, whose ongoing critique, queries and suggestions gave form to the vision behind this book.

Contents

It is only with the heart that one can see rightly;
what is essential is invisible to the eye.
— Antoine de Saint-Exupery

Everything the Power of the World
does is done in a circle.... The life of a man
is a circle from childhood to childhood.
And so it is in everything where power moves.
— Black Elk, Oglala Sioux (1863–1950)

Preface

This book is written in a way that allows the reader to enjoy the stories individually or in sequence, as part of a larger narrative. The stories themselves have grown out of the experiences that have been closest to my heart. They are primarily inspired by the dolphins who by various quirky turns of fate found their way into my life, bringing with them exceptional magic and profound joy. They are about how the dolphins in their own enigmatic way have become teachers and guides — always full of humor and enduring grace. Both in and out of the water, they have come into my awareness carrying messages and inspiration.

The narrative is interwoven with the fabric of family and friends and other animal companions. We are all connected by invisible threads — life's intelligence is inextricably interdependent, and our thoughts and feelings appear to influence each other and our reality. We are co-creators, fashioning ourselves and our beloveds into our own life stories.

The tellings here are about love. They are about realizing the depth of our heart's affections, and then having to let go. They are about the frailty of our human existence and our own vulnerability. The stories also point to something beyond these perceived limitations. They reveal a universe that is infinitely supportive, if only we would open ourselves to receiving what is continually being offered. It is knowing that what we may call God, or our connection to Source, *is* Love itself, and is something that is with us always. We travel the world searching, looking for answers or for a particular person, only to return home and find that love has been at our own doorstep, patiently waiting all along.

CHAPTER 1

From
Delphi to Dolphins

Diviner than the dolphin is nothing yet created...
—Oppian of Silica, *Halieutica* (200 A.D.)

The tantalizing image of a small dolphin leaping through curling waves is like seeing a prism refract a beam of light out of the corner of your eye. One wonders — what *was* that? A blithe spirit playfully riding the rhythms of the sea? God must have been having a very good day when he (she) created such a creature, with its bold, irresistible smile, the simple curve of its body, the wide and luminous eyes. This watery trickster teases us out of our doldrums and reminds us to smile and be happy. Enjoy. Be joy. Perhaps life isn't what we *think* it is. Perhaps it just *is* — a never-ending dance upon the changing tides.

As an avid reader and student of art history and mythology, I had seen the painted images of dolphins leaping from the foam-capped waters of the Aegean Sea.

They encircled Greek vases and appeared on temple reliefs, and they floated across the pastel murals of ancient Crete. These pictures radiated enormous freedom and peace — their synchronized movements evoking sublime beauty, infinite energy and a sense of surrender. What lay behind that elusive presence and mysterious smile? What was the truth behind the *mythos* and the legends?

What began for me as a curiosity took hold in my imagination and grew. Eventually it came to captivate my heart, and like other love stories, led me into all kinds of unanticipated adventures and at least as many awakenings.

My childhood was filled with travel and the adventure of new places. When my father was a young U.S. army officer stationed in Germany, he was given a temporary assignment in Athens, Greece. The four months we spent there contain the most vivid of these early recollections. We lived in a crumbling villa on the outskirts of town with an old gnarled olive tree in the back. I used to hide up in its branches, stuffing myself with its tiny black fruits. What I remember most is the warm light everywhere reflecting on white — whitewashed stucco walls, clean cotton shirts and sheets fluttering in the sun, the pale sunbaked earth beneath our feet, and the bleached white shells and barnacles sticking to the hulls of painted wooden boats that had been dragged up onto pebbly beaches. We visited many of the ruins and temples — the Acropolis, Corinth, Olympia, Mycenae, the Temple of Poseidon and Delphi — their timeless stones honey-colored and serene. With their simple and elegant beauty, these columned structures, poised perfectly between heaven and earth, conveyed a sense of balance and time-honored proportion. They also appeared shrouded in mystery, as if the gods, when called upon, could still speak through them. Later, I would come to associate the ancient Delphic wisdom with the graceful and enigmatic dolphin.

My mother at the temple ruins of Delphi.

The Greek islands were another source of delight — their small villages clinging to the worn rocks; colorful *tavernas* with the smell of fresh bread and carafes of wine set on rickety round tables; shaggy donkeys with clumsy loads tied onto their backs; the noise of seagulls; and children running with bare feet. We often swam in the sea. Every day it seemed we were in the water, the sun and salt bleaching the color from our hair, while an endless array of exotic fish and octopi pulsated around us. This translucent tapestry of water and light and the ancient edifices that adorned the arid Greek landscape became my first experience of what I would call spirituality.

Hydra Island, Greece.

These early memories — of the islands, the temples and the sea — flooded my mind many years later, as I stood and stared at the illustration of a dolphin that had fallen from a bookstore shelf. It was 1994, and I was looking for inspiration — a name or a guide — for my new graphic design business.

I had moved to Northern California and was living in Berkeley, and in my spare time had become interested in computers. I was fascinated with the idea that images could be placed inside a computer, altered, manipulated and neatly blended with words and story. I was hooked, and I decided to bring my various artistic inclinations to bear on modern cyberspace and become an entrepreneur. The synchronicity of the small dolphin virtually dropping into my hand sparked my imagination. I knew the dolphin embodied qualities that I wanted to emulate in my art and design work. The dolphin was a mercurial and magical being, and like the ancient Greek oracle of Delphi, seemed able to tune in and receive images or messages. It seemed an appropriate symbol for wanting to render ideas into beautiful form. I later learned that the word dolphin has the same root as Delphi, and means *womb,* or source of all life and wisdom. The name of my venture became Dolphin Press, and it was the beginning of a whole new cycle in my life.

෴

A few years later I was still holding my own as an independent businesswoman, but an event was to occur that would cause me once again to reevaluate what it was that was truly important to me and the direction and purpose of my life. I was to experience in no uncertain terms the fragility and temporalness of our journey here.

CHAPTER 2

Close Encounter

If you would indeed behold the spirit of death, open your heart wide unto the body of life. For life and death are one, even as the river and the sea are one.

—Kahlil Gibran, *The Prophet*

I opened my eyes, groggily, blinded by the glare of harsh white lights. I lay on my back as I was rolled down a long hallway. I was thirsty. I tried speaking, but my voice sounded gravelly and harsh.

I was very tired and drifted back into an uneasy sleep.

Later I awoke in a strange room and looked at a clock on the wall. 3:30. 3:30 in the morning? The light had been switched on, and a nurse came to stand at the side of my bed. "Just checking your vitals," she said in a comforting tone. I saw an array of tubes taped to my arm that were connected to an elaborate machine on my right. I was grateful when the lights were dimmed and I could drift away again.

The next morning a doctor appeared in my room. He came to sit at my bedside and looked at me with a faintly quizzical expression. "You gave me quite a scare last night," he said. "You're lucky to be here."

Vaguely, I remembered him as one of the doctors from the emergency room the night before. Friends had rushed me through heavy traffic to Marin General Hospital. I had been bleeding — hemorrhaging, we later learned — from a tumor in my womb. I had fainted. In the car a girlfriend kept talking to me. "Stay awake," she said. "Stay awake." Later, as medics rushed to give me a blood transfusion, I remember feeling cold. I had started to shake and asked for a blanket. Strangely, at no time had I been afraid. Inwardly I felt calm, even though the frustration and fear was apparent in the faces around me. Time had seemingly slowed down, and I felt oddly disconnected from my body. Despite the panic in the room around me, I felt protected by some dimly perceived, expanded other presence that assured me that, whatever was happening, I was safe, and everything was going to be all right. I had only a faint memory of being hurried into surgery, of a mask covering my nose, and of being asked to slowly count backwards as I was put under anesthesia.

"That was quite a night," my doctor continued. "About five more minutes and we would have lost you. We had to perform a hysterectomy. The good news is you still have one set of ovaries."

Quietly I digested this new piece of information. No children. I think I had already let go of that possibility. Somehow all I could feel was relief and the surprise of being here at all. "Thank you. Thank you for being the one on duty," was all I could manage.

Suddenly, I found that my life had been reduced to performing very simple tasks. I had no sensation in the lower part of my body. The aftereffects of the anesthesia lingered, and the constant drip of various fluids into my veins kept me sleepy and sluggish. Before I could eat real food, I needed to regain the use of my bowels. My goal became getting to the toilet. I advanced to taking short walks in the hallways, my tubes and apparatus dragging behind me. I flopped along in my one-size-too-big hospital slippers, trying to keep my short cotton gown from flapping open.

I clearly saw the humor in my situation. In fact, I felt remarkably light in spirit, as if completely removed from the complexities and challenges of daily living and human interaction. I was now living inside a rarified bubble where everything had been simplified. All the while, I continued to feel the support of an overriding other presence.

If I do enough laps, perhaps I'll be rewarded with a meal, I thought. The day arrived when a tray was ceremoniously brought in to where I lay, propped up against the pillows. My dismay must have shown on my face. "Is everything all right?" asked the nurse. A cup of coffee and a bowl of raspberry jello stared me in the face. I grimaced.

The next time the phone rang, I was relieved to hear the voice of my acupuncturist girlfriend Nan on the other end. "Nan, I need real food," I whispered urgently. "Could you bring me a smoothie when you come to visit?" She did, along with some restorative Chinese herbs that I hid in the cabinet next to my bed.

A few days later, I asked the nurse if I could see my doctor. When he arrived, I shot him a conspiratorial look. "Do you think you could give me permission to go home? I'm actually feeling okay." I added, "I haven't had a real meal since I got here."

He looked at me appraisingly. "Sounds good to me. But before you go," he laughed, "I'll have them order you up something from the kitchen." I left the hospital in my nightie and slippers and with a full stomach of mashed potatoes, turkey and gravy.

My physical trauma notwithstanding, I was ready and eager to fully re-engage in life. Within a week I was out dancing, albeit gingerly, and within three I backpacked in Yosemite. My doctor gave me the thumbs-up. "You're my miracle," he said.

By the fall I was back to full-time work, but what I had become acutely aware of was the frailty and transitoriness of life. I felt that I'd been given a second chance. I'd been taken to a door that most people, when they get to, are required to walk through. I was certain that if I had gone through that door I would have entered into a realm of unimaginable peace and light.

Throughout my time in the hospital it was as if I had been protected, watched over, by something greater than could be explained by what was happening on a physical level. I found myself asking, if life has the potential to end so abruptly, and if death is not something horrible and dark and final, then what keeps us from doing more of what we really want while we are here? What am I afraid of? What's to keep me from living life more fully now? What *is* really important to me?

⌒⌒⌒

As the new leaves began to appear on the trees in the spring of the following year I decided to follow some of my dreams. One weekend, my curiosity piqued by an inviting lecture title, I found myself in a room full of people waiting to hear a presentation on dolphins. The event took place at the Whole Life Expo in San Francisco, a huge gathering consisting of workshops and product displays involving everything from the latest in consciousness-raising techniques to holistic health — a sure setting for something out of the ordinary to happen.

When the speaker arrived, we were given a brief introduction, and then the audience was invited to close their eyes and to join in a group meditation. We were asked to pay attention to the breath, to gradually quiet the mind, to let go of the stimuli from the day — the lists, the worries, the conversations — and to sink into a place deep within. *Feel the weight of the body sitting in your chair. Feel yourself relaxing. Allow the tension in the muscles to drain out, letting it run down into the floor.* The voice of the speaker continued hypnotically, gently guiding us. Slowly, I felt my breath deepen. I could feel the weight and the drowsiness in my body, and then an ever-so-slight twitching in my eyelids. I knew I was entering into a light trance. I began to feel myself floating slowly down through space.

As I sank, I gradually became aware of the watery blue light of the ocean all around me. Endlessly it stretched, expanding in every direction, and like a diaphanous angel I continued to fall, gliding into its depths. Long before anything else, I heard a soft whistling. With it came an almost imperceptible presence — radiant and softly inviting. Out of the calm and now inky darkness I saw two shadowy forms, moving gently and certainly towards me. The translucent shapes of two dolphins began to reveal themselves, shimmering and winking in the near blackness. Slowly, they began to spiral around me. As they spun their silver rings, the space inside my chest began to ache, and the pounding of my heart erupted with profound and indescribable joy. What I heard was, *You are so loved. You are a precious being.* I felt an explosion of light inside me, and I began sobbing. I felt overwhelmed and completely taken by surprise. What was happening? In this silky, watery world, the dolphins' presence had touched me at a place in the center of my being — a place of inexpressible love and joy and beauty. I knew I was being given an exquisite gift, an experience of the core Self — that which I Am. And I knew without a doubt that this was Home — a place of profound interconnectedness. Again I heard, *You are This.*

As we were gently guided back into the room and taken out of trance, I thought about my two underwater messengers. I knew they had activated what could only be called a core or transcendental experience. Why and how had they become such powerful messengers? I knew I wanted to encounter dolphins in three-dimensional reality. Aside from always having been intrigued by them, I wanted to see and learn for myself who and what they really were.

CHAPTER 3

Letting Go

*Faith means living with uncertainty — feeling your way
through life, letting your heart guide you like a lantern in the
dark.*

—Dan Millman, *The Laws of Spirit*

I started researching dolphin swim programs and trips.
Later that summer, I attended an engagement party for
friends where I met Jon, a personal growth and work-
shop leader. He was leading a trip to Hawaii the following March to swim with the wild spinner dolphins on
the Big Island of Hawaii. Our mutual interests soon
sparked into romance. I signed up for the trip and agreed
to help design the visuals for his flyers and advertising. I
was thrilled! In my imagination I immersed myself in the
dancing waters of Kealakekua Bay, I flew with the
Goddess Pele over rivers of underground molten lava and
sacred caves, and I felt the breath of balmy ocean breezes
on my skin.

As the months went by and we got closer to our
departure date, I began to wonder what the chances
were, realistically, of actually finding the dolphins. After

all, we were meeting them somewhere out in one of the largest natural bays in the Hawaiian islands. I felt a great sadness well up inside me as I considered the possibility that this encounter might not happen. For days I struggled within myself, wanting to prepare myself for a very real scenario — the likelihood that they would not be there. Over and over I've observed this dilemma between the doubting mind and the heart. The heart longs and aches, and the mind scrambles to protect us from disappointment, from failure, from disillusionment. For days I prayed and had conversations with the dolphins in my head. Finally, I came to a place of letting go. I let go of my attachment to seeing them. If they chose not to come, that was okay. I would still enjoy my vacation in Hawaii. Nothing would be lost. In fact, everything would be perfect just as it was.

It was at this place of detachment, of letting go and surrender, that something miraculous happened. I was very busy with work the week before we were scheduled to leave. I was putting in long hours, and I had countless details to attend to. Then, in the midst of all this preoccupation and noise, I started to hear something else. I started to hear, faintly at first and then louder, small distinct chirpings and whistlings. It became unmistakable — it was the sound of dolphins, and it got louder. I don't believe this, I thought. I signaled back anyway: Thank you for communicating, but now I'm having a hard time concentrating. All week long it was like being tuned into a very special and exclusive radio frequency.

At the end of the week we flew from San Francisco to the town of Kona on Hawaii. From the air I could see the moonlike lava landscape of the west shore. We arrived at our beautifully situated hotel south of town, ate dinner and then headed for bed. We were scheduled to wake up early, at 5 A.M. the next morning, to carpool to Kealakekua Bay with our wetsuits and snorkel gear. In the haziness of early morning light we sheepishly greeted one another, coffee cups in hand. My heart hammered in my throat. The moment had arrived. Would the dolphins show up for their date — an invitation made through the ether and precipitated in the heart? Slowly, we drove the winding road down towards the glistening waters of the bay and pulled into a sandy parking lot. Large red hibicus flowers lay strewn across the ground. I walked toward the beach, and then I saw it — the splash of a single dolphin jumping just off shore.

Karin and Jon at the City of Refuge on the Kona coast of Hawaii.

I was so astonished that I started to cry. I realized then that if this was to be the only contact we had with the dolphins all week, I would still be extraordinarily happy.

To me, they had decided to keep our date. Later, at the end of our swim, I spoke briefly with an old Hawaiian man who sat watching our foray out into the water. He grinned and quietly commented, "They haven't been here for weeks, but today they are here."

Spinner dolphins off the Hawaiian Islands.

We did find a huge pod of dolphins — or perhaps they found us. They showed up on each of the three days we had hoped to swim with them. It was magical and extraordinarily dreamlike — like being in an altered state of reality or another dimension. In the evening Jon led us in meditations. "Visualize," he said. "What more do you want to create for yourself with the dolphins?" My inner vision had been flooded with brilliantly colored pictures of the dolphins ever since our first swim in the water. It was like watching my own inner nonstop movie. I wondered — was I creating the pictures, or were the dolphins sending them to me? As I sat quietly attending to the in and out of my breath, I saw myself gazing into the eye of a dolphin as it slowly swam next to me. Then another one leapt high up in front of me. The next morning I found myself transfixed by the gaze of a dolphin as he gracefully swam past me. Then a loud splash caught my attention as a dolphin leapt into the air, spraying me with water. I laughed out loud. They must have gotten my message. Or maybe I got theirs.

Mandala of shells, flowers and leaves.

One of the things we discovered was that the dolphins liked playing a game with leaves. They particularly seemed to like the large yellow leaves that floated out from shore. The dolphins would pass them from one fin to another, sometimes catching them on their flukes (tails) or carrying them on their rostrums (their long beaklike jaws). As a group we decided that we would come down to the bay for a fourth day and bring the dolphins a gift of leaves and flowers. That last morning we carefully swam out with our gifts, looking for the dolphins, but they had disappeared. We had not had a prior agreement to swim with them, and in their enig-matic fashion they had quietly vanished. We returned to the beach, and on the sand we created a farewell mandala of shells, red hibiscus flower petals and yellow leaves. I was touched by the delicate beauty of our fragile cre-ation. It seemed appropriate that our last encounter would be with our group together standing in a circle holding hands, with the temporal beauty of nature spread out at our feet.

The teaching for me here was about the power of the heart, of letting go and surrendering. What I learned was that we are enormously powerful if we choose to create out of love. I felt as though I understood in a new way the old adage: Let go, and love will find you. With the distance of hindsight, my mind would sometimes argue that I tend to have a very overactive imagination and that I am a prime candidate for hearing and seeing things. How would I ever know whether the dolphins would have shown up regardless of anything I did or felt? Wasn't it all just a matter of random chance?

Over the years, however, the truth of these experi-ences has become more and more palpable. The dolphins are a constant reminder to check in with myself and to ask whether a wish for something or someone is truly coming from my heart. If the answer is *yes,* then those things — be they people, projects, places, experiences — seem to come towards me. They *do* show up. It is not a logical road. If my desire is coming from a place of ego or of trying to control a situation, the outcome is less

predictable. The "message" doesn't seem to get through
— or if it does, it doesn't seem to have much power.
Over and over again, I have heard the communications
to *relax* and *surrender* and *let go*. As a result, I find that I
cry more, and I laugh more. I try to let myself be in the
river of life, no matter how scary it may sometimes
appear — to go with the flow rather than trying to resist
it. I try following my intuition or my gut, often down a
seemingly illogical path. Ultimately, the power of Love
seems to find a way. It appears to be irresistable.

Spinner dolphins close-up.

Since these initial experiences, I have become familiar with the term *telempathy,* a phrase coined by Joan
Ocean, who has spent years swimming with the wild
spinners in Hawaii. Telempathy is a combination of tele-
pathic and empathetic communication, or empathy at
a distance. Empathic communication occurs when we
experience the exact sensations of someone or something else with whom we are emotionally close. My
own experience has shown me that dolphins tend to be

extremely empathic. They seem to have the ability to feel the pain and emotional state of another being. This, combined with their echolocation or imaging skills — the ability to project clicking sounds (created in the air sacs beneath the blowhole) out in front of them, then interpret the soundwaves as they are reflected back, thereby determining the size and distance of foreign objects — seems to make for a very sophisticated form of telepathy. I am reminded of a woman in our group on our trip to Hawaii who was pregnant. She didn't go into the water for the first couple of days because she felt tired from the flight. When she finally did, she was surrounded by dolphins who seemed to show a particular interest in her. It was if they knew she was carrying a child and needed special attention. The combination of these two skills — the ability to be empathic and also to "see through things" — makes the dolphins especially suited as "healers" (by their very presence) and as messengers, perhaps even cosmic messengers.

When people ask if dolphins have changed me, I say that I seem to have more dreams now and fewer plans than I used to. I hold my dreams out in front of me and then let them go. Invariably my dreams show up in unexpected ways and sometimes in new forms — *here we are, it's time, here's the connection or the opportunity.* I worry less about the details and spend more time putting color into my daydreams, adding scents and enjoying the warmth of the sun on my skin.

In the midst of great change or loss I am reminded to trust that everything is unfolding perfectly. Stay calm, listen and catch the next wave. I try to practice living in dolphin time. To me dolphins live in circular time as opposed to linear time. For many of us life appears to move in straight lines, but perhaps it is more accurate to say it moves in many directions at once perfectly synchronized. We are not separate from one another, but part of a much greater pod that has its own intelligence. Our job is just to tune in and then get out of our own way.

CHAPTER 4

Copan

> *[Dolphin] became the carrier of messages of our progress. The Dreamtime dwellers were curious about the children of Earth, and wanted us to grow to be at one with Great Spirit. Dolphin was to be the link.*
>
> —Jamie Sams and David Carson, *Medicine Cards*

Nine months after returning from Hawaii, on a rainy Northern California winter day, I got a call from my sister in Seattle. She sounded unusually upbeat. "What's up?" I asked.

"I just got back from diving," she said. She had recently returned from her favorite place on earth, an out-of-the-way dive resort in the Bay Islands in the Caribbean. "And guess what?" she continued. "They have dolphins."

That got my attention. Ever since my trip to Hawaii I had dreamed of having more intimate contact with the dolphins. Was it possible to play more intimately, one-on-one, with them? I knew that the bottlenose dolphins were highly social and had been known on occasion to actually seek out contact with human beings. Also, since swimming in the ocean off Hawaii, I'd had the desire to go someplace where the water was *really* warm, where wearing a wetsuit was not a requirement.

A week later it was still raining, and I was on my way to renew my passport. I had to twist Jon's arm, but finally he gave in. "You mean it's just a regular resort?" he queried. "No workshops?"

"No workshops," I replied firmly. "I promise. This will be good for you."

≈≈≈

I slipped off the platform into the warm, silky water. It was March in the Caribbean, and I felt like I had entered into an endless summer. Tilting my head back, I floated, enjoying the sensation of being submerged in a fluid world. I could smell and taste the sea salt as my arms and legs lazily pushed against a gentle current. I felt like I'd come home. While growing up, I had sailed with my family, and like my sister, later I learned to scuba dive. All my life I'd had a nagging compulsion to be near water, but now as I peered across these blue depths expectantly, I felt the first sensation of butterflies in the pit of my stomach.

Jon and I had signed up for the Dolphin Specialty Course, a small program being offered at the Marine Biology Institute on the resort premises. It included a couple of lectures, some hands-on demonstrations, and dorsal tow rides and foot pushes with the dolphins. Today was our first session in the water.

The resort and locale exceeded our expectations. It was paradise — refreshingly authentic with large flowering foliage, wide walkways and comfortable cabanas for the guests. Two small tropical keys floated in a wide bay. A pod, or extended family, of about sixteen bottlenose dolphins lived in a large enclosure off one of the keys. A couple, both marine biologists, ran a center with a small staff of trainers, most of whom were from the island. I had never experienced dolphins in this kind of setting before, so my main question was, were the dolphins happy? It appeared as if they were. They lived in an offshore marine habitat, much like the one they had come from in the wild, and they received top-level attention and care. What impressed me most was the rapport between the dolphins and their trainers. A feeling of genuine warmth and enthusiasm permeated all of their interactions.

Earlier in the week we had attended several classroom sessions, which included video footage focusing on the history and physiology of dolphins and dolphin research. Later we sat on one of the training platforms attached to the catwalk that surrounded the training area for the dolphins while they showed off their sonar and echolocation skills. This morning had been spent in a hands-on anatomy lesson. We had joined one of the biologists at a platform and, at the blow of her whistle, a dolphin had swum up alongside us. It had proudly shown us a full mouth of formidable teeth, a powerfully agile fluke (or tail), and the small blowhole on the top of its head. A favorite part of the lesson for the dolphin was being touched. We stroked its unbelievably smooth, white underbelly, and even its tongue.

Now we were being invited to join some of the dolphins in the water for a closer encounter — and for our rides. I began to swim out to the far end of the training area. It dawned on me that this was the closest I'd ever come to a large mammal and that I was swimming directly towards it. We had been told not to make any sudden or erratic movements, to keep the tone of our voice low, and to adopt an air of confidence and composure. I wondered about some of the finer points of dolphin etiquette. What was the proper way to greet a dolphin? Would the dolphin be as interested in me as I was in the dolphin? A dark, shiny head emerged from the water, and I could see the characteristic dolphin grin traced along its jaw as he beamed out at me. He was huge! He lifted his chin. Without thinking, I instinctively reached out to stroke his neck as I would my cat. But what got my attention was the look in the dolphin's eye. He peered straight at me with a look that was both wise and amused. *Welcome and thank you for coming* seemed to be the message. I immediately had the revelation that this guy ran the show! Suddenly the trainer's whistle blew. *Gotta go — got work to do.* In that moment I knew that this dolphin was a leader — a kind of ambassador — and that he knew exactly what he was doing. His job was to serve as a link between his species and ours, providing a bridge between the world of humans and dolphins. Many of us will be ambassadors, I thought. The world is changing and all of its life forms are in need of assistance. We are taking on new roles and opening up to new ways of communicating, even across species. There is much we need to learn from one another.

Even in the warm water, I could feel goose bumps on my skin. It was as if my mind had been read, and I had just downloaded a big chunk of information. Ever since we had arrived, I had wondered whether this could be a place where Jon and I could bring people for a deeper and more intimate experience with dolphins. Now I felt as though I had just gotten an answer in the affirmative.

My encounter session with the dolphins continued. Two sleek bodies swam up on either side of me. I took hold of a dorsal fin in each hand, as I had been instructed, and flew, laughing, through the water in a wild cloud of spray. Later I learned from the trainers that the name of my "ambassador" was Copan (Koh′paN),★ which means king in Spanish. Copan was the largest dolphin of the pod. He was the alpha male.

⌒⌒⌒

A year passed from the inception of our idea to the realization of a dream. After countless e-mails and numerous negotiations, and with the generous assistance of the marine biologists on site, Jon and I launched the Dolphin Trainer for a Week program. It was designed to provide both dolphins and humans a chance to get to know one another and their vastly different worlds in an intimate way. Today our human/dolphin program has grown to include people from all across the United States, Europe and Canada. It has become a kind of school — a place of profound learning and play.

★The N takes the nasal sound of the preceding vowel.

CHAPTER 5

Little Guru

Death is not extinguishing the light; it is putting out the lamp because dawn has come.

—Rabindranath Tagore

As my connections with the dolphins evolved, another incident took place which was to remind me again of life's impermanence and how some of the more important communications we receive in life can come from surprising places. I have always been an animal lover, and my animal companion for seventeen years had been a little black and white cat. Her start in life had been shaky. She had grown up as the runt in a neighborhood of toms and had her first litter of kittens shortly after her first birthday. Although she belonged to a neighbor, I would often find her hiding out on my back porch where she would sneak in through a tear in the screen door.

One day the doorbell rang. It was my neighbor, with a request. She was moving, and they couldn't bring two cats with them to their new home. Would I consider taking Blackfire — that was the cat's name — until they could find a permanent owner? I didn't have to think twice. Weeks stretched into months, and I finally I asked to keep her. Over the years I watched her evolve from a skittish scaredy-cat to a wise and gracious grandmother who loved being the center of attention.

☙☙☙

Just before Thanksgiving, as my brother had concluded a visit with me and was driving away, my eyes caught sight of Blackfire as she came into the room. Was it my imagination, or did she seem a little more low-key than usual, listless even? I made a mental note to give her some extra attention.

The next day, as she crossed the living room floor, I noticed a fine weave to her gait. "Is everything all right?" I asked, now worried.

The next morning I woke up with a start. It was still dark outside. I watched as Blackfire moved unsteadily towards me. Something's definitely wrong, I thought. I'm taking her to the vet.

"At this hour?" Jon grumbled. "It's the holidays!"

"There must be an emergency room clinic somewhere," I insisted. I rummaged through the phone book.

We drove Blackfire about five blocks to a 24-hour veterinary emergency clinic. There we sat in the waiting room as the doctor took a look at her and ran some tests. When he came out from his office I could tell by his face that the news was not good.

"I'm sorry," he said. "She has kidney failure. I've put

some fluids into her system temporarily, but there's really nothing you can do. Her numbers are off the chart. This has been something that has gradually gotten worse over time. You'll have to put her down."

I looked at him in shock. "Put her down? There must be *something* we can do." I turned to look at Jon and I saw tears starting in his eyes. He used to say that Blackfire was the sweetest being he'd ever met. I knew this was hurting him as much as it was me.

"No, I'll take her home," I said. I bundled her up in her blanket and we drove home. I was not ready for this, I kept saying to myself. She was such a big part of my life that I couldn't imagine living without her. I knew that in Chinese medicine the kidneys were related to fear. Was her early childhood catching up with her? I felt incapable of the decision to end her life. I couldn't believe that this was happening so quickly.

At 3:30 in the afternoon Jon called. He'd gone home to take care of some work and then had started making calls. "I want you to take Blackfire in to see another vet," he said. "I'm coming right over, and I'm going with you." He'd found another clinic in San Rafael. "He's willing to see us for a few minutes just before closing at 5 o'clock. He's kind of alternative," he said. I had the results of Blackfire's tests faxed over to his office.

"Well, is there any hope?" I asked the new vet. "Are her numbers that bad?" He ran his fingers through his hair.

"Oh, I think I've seen worse," he said. "Here, come with me. I need to show you a few things. If you're going to keep her alive, she's going to need daily transfusions of water injected right under her skin. I'm going to show you how to do that."

We watched as he hung up a plastic bag of fluids, a plain saline solution, with a long tube attached to it. "You're going to need to use a new needle every time." He quickly inserted a sterile needle into the attachment at the end of the tubing, released the flow valve, and then expertly slipped the needle into the loose skin at the nape of her neck. "See the lines drawn here on the bag? They mark how much water she's going to need. She's going to require two transfusions a day. I'll also give her some homeopathic remedies to help stabilize her." He explained that she was going to get picky about her food and that she needed a balanced diet low in protein. He would need to see her every couple of days for awhile. Maybe the water therapy would help bring her around and get her numbers down.

For two weeks after that I drove to the vet with her in my lap almost every day. Most of those days I was terrified, and I cried uncontrollably. I thought I knew something about death; I'd been five minutes away from my own. Why was this so different? Please let her numbers come down. When they didn't, I started praying. I'm just not ready for this, I kept saying. I need more time. We became a familiar sight in the clinic.

In amazement, I noticed that Blackfire would come and sit down on her spot on the bed when it was time for her daily transfusions. Quietly she purred as I gently stuck her with the needle. *It's okay* was what I heard. *Everything is going to be all right.* She seemed to know that I was the one who needed reassuring. The vet thought she might live for three more months. I decided to make the most of it.

Blackfire lived for another year. Her water treatments became part of our daily routine, as did her other remedies, which I would slip under her tongue or give with an eye dropper. When a year was over, the day came when Blackfire restlessly shrugged away the needle. I could feel the panic rising up in me.

"If you don't have water, honey, you'll die." I was torn about what to do. She avoided my room and hid in a closet in one of the other bedrooms. I felt her frustration and depression, so I hired an animal communicator (someone trained to telepathically communicate with animals). It seemed like her time had come, but I wanted another opinion to make sure I understood what Blackfire needed and that I would be doing the right thing. When it came time for our phone conversation, I brought Blackfire in to sit at her favorite place on the bed.

The communicator tuned into Blackfire and her energy field, and then she began speaking to me. "She's tired, you know," she said. "She's ready to go. She's had a great and long life with you. She wants you to stop with the water. She doesn't need or want to be put down. She wants to stay with you till the end. Don't worry about her. She knows how to die. She's done it many times before. She's very pleased and grateful to have been able to share this time with you." Everything she said resonated with what I already knew. She said a few more things, and then I hung up the phone. Blackfire was purring contentedly.

"All right," I said, "No more water, and we're doing this together." It was Thanksgiving again, and I would stay with her.

For the next five days Blackfire stayed in her spot, except for a few times when she wobbled to her litter box. Never once did she have an accident. At night she slept by my side. Gradually, I saw the light go out of her eyes. I felt like I could actually see a faint haze leaving from the top of her head. She would lie on her side breathing quietly. All her energies seemed to be focused on letting go of her body. *This is a great teaching for you,* I heard. I want to be with you when you go, I pleaded. Saturday night on the weekend after Thanksgiving Day I awoke in the early hours of the morning. Everything was strangely still. An eerie silence surrounded me. I reached my hand over to Blackfire, as I had gotten accustomed, to see how she was doing. Her body was stiff. I realized that she'd left and was no longer in her physical form. I could see a pale, diffused light filling the room, and I knew in that moment that she was everywhere. She had merged with everything, and was closer to me than she'd ever been. I felt as though I'd slipped through the veil and for a moment hovered between the worlds.

The next day I laid her body on the floor, covering it with her blanket. I had always been attached to her

sweet form, but now I saw her body for what it was —
a worn-out coat. Blackfire was gone. I had been unde-
cided about what to do with the body. I wasn't sure I
felt comfortable with burying it. Now I knew I would
have it cremated. With her body returned to the elements, she could perhaps transition to spirit more
easily. All day long I felt waves of overwhelming grief
even after a year of preparation. I felt my aloneness and
the silence.

On Monday I left her body at the vet. I couldn't
speak as I laid her on the counter wrapped in the
familiar blue blanket. I could see the tears in the eyes
of the people who had known her. She was well loved.
Then I got into my car to drive over the Richmond
Bridge to the East Bay for an appointment.

About half way across I suddenly had an image of
Blackfire in my head. She was laughing. She had little
wings coming out of the back of her neck and she was
flying. She looked so comical I had to laugh too. *Thank*

Blackfire.

you, she said. *Everything's all right. I'm sorry I left while you were sleeping, but it was so hard to detach completely when you were awake and pulling on me. It was easier to go while you slept.*

Yes, that makes sense, I thought. You have been my little guru. No one else could have initiated me into one of life's greatest mysteries better than you. Fly high, little girl. Thank you, and I love you.

CHAPTER 6

Misha

There are only two ways to live your life. One is as though nothing is a miracle. The other is as if everything is.

—Albert Einstein

Springtime is a good time for new beginnings, or so I rationalized. Only a few more months, three at the most, and I would get a new kitten — someone to fill the void Blackfire had left. I reassured myself of this plan as the windshield wipers kept steady pace with another downpour. It was almost New Year's, and I had some free time on my hands. Perhaps I could go and just look at some kittens.

In the following week I visited three animal shelters before calling the hotline for the Marin Humane Society. A recorded announcement told the listener which dogs and cats were up for adoption. "One male kitten from a litter of three. Part Siamese. His name is Luke," the message said. Luke, I said to myself, and I immediately thought of Luke Skywalker from *Star Wars*. If he takes after his namesake, perhaps he's a little Jedi.

When I found Luke he was by himself in a small wire cage. He had been separated from his brother and sister and had arrived a day later. He was all gray, with soft medium-length hair, a large fluffy tail and round, intelligent eyes which stared straight at me. I could see the Siamese influence in his paws and ears, which were highlighted with pale silver. He had a strong, handsome face. He allowed me to pick him up, and I gently held him.

"If I didn't know you were a cat, with that color I'd say you were a dolphin. Who'd have thought I'd end up with a gray cat."

Spring or no spring, Luke came home with me.

$$\backsim\backsim\backsim$$

Although I liked the name Luke, another name kept coming to mind as I got to know my new companion. The name was Misha. I liked the softer sounds, the lilt of two syllables. On the Internet I came across several photographs of cats with similar coloring and markings. They were Russian Blues — all the more reason for him to have a Russian name. Misha is short for Mikhail or Michael, meaning *who is like God* in Hebrew. Luke comes from the Greek meaning *bringer of Light*. Either way, I could see that my new cat was destined for greatness!

In the weeks that followed I saw the world through new eyes — the eyes of wonder. The bathtub full of water, the candle flame, the army of ants crossing the kitchen floor — Misha would sit quietly watching these for hours. And of course he liked to play. There was the string monster and the mousie and his favorite — the crinkle toy. The crinkle toy is simply a crumpled-up piece of aluminum foil crunched into a light ball. I would throw the crinkle toy and Misha would dash madly across the room in happy pursuit. What surprised me was that Misha would then bring it back. He loved playing fetch. "You must have been a dog in your past life!" I said.

He was also funny, and would often try to amuse me. I would find crinkle toys in my shoes or on my pillow when I went to bed at night.

When Misha was five months old he went through his wild stage. He was everywhere. He would streak along the back of the couch, then across the coffee table to the armchair and back, his claws raking along the upholstery. My housemate came to me with an ultimatum: me or Misha — one of us has to go, or change. Misha had a scratching post, but so far, the post was apparently no different from any other piece of furniture in terms of places to scratch. Clipping his nails had helped, but it had not solved the problem. I did not want to declaw him, so I did some research, and came home from the pet store with a small bag in my hand. You and I need to talk, I told Misha.

I explained to him that scratching the furniture was not an option, and I reminded him again of his scratching post. He watched attentively as I opened my bag and showed him my new purchase — Soft Claws. Soft Claws are miniature plastic adhesive mittens designed to slip over a cat's nails. As I explained this, Misha looked at me with a serious expression on his face. After that everything changed.

To this day Misha scratches his post, and the Soft Claws have stayed in the closet.

The house I live in has a wonderful deck off the back which is enclosed by a high wooden fence and lattice covered with climbing honeysuckle, clematis and jasmine. For the first few months I would let Misha out and he would happily play, not having figured out how to get over the fence. Then he became curious, and I knew there would be no turning back. I felt like a protective mom. Blackfire had grown up streetwise; she had always roamed, and I never worried about her. Only once in all the years that I had her did she get lost, and that was for a brief three days. Misha, on the other hand, seemed so naive and innocent I was reluctant to let him explore the outside world. At the same time, I did not want to deny him the experience of more freedom. I decided to make a deal with him.

"When you are older and bigger — one year old," I said to him, "you can go over the fence." By the end of eleven months, I knew I had to keep my promise. Again and again I would find Misha tangled up in all the jasmine, gazing out at the big wide world. Finally, I cut a little hole in the lattice and let him climb through. The first time he left he was gone for over twenty-four hours. I combed the neighborhood calling his name and searching with a flashlight late into the evening. At 2 A.M. he bolted through the door. He'd obviously had a big adventure. His eyes were wide and his hair unkempt. The next day I discovered a small wound in the middle of his back. I washed it with hydrogen peroxide and let him go back to sleep.

Misha.

I did not close the hole in the fence, and when Misha recovered, he continued his explorations. Now, a year later, Misha spends most of his time under the lavender in the flowerbox outside my bedroom door. He's not a roamer. He's a homebody, and softer and more gentle than ever.

CHAPTER 7

Heaven Within

Thus my eyes have filled with warm
Soft oceans of divine music,
Where jeweled dolphins dance,
Then leap into this world.
—Hafiz, *The Subject Tonight Is Love*

There are certain places on the planet that are particularly enchanting and seem to possess, to an unusual degree, the power to heal, restore and rejuvenate. For me, one of these is Maui, an island that had made an indelible impression on me when I'd first come to visit almost twenty years ago. At the time, I had fallen in love with its peaceful green hills, the waves as they crashed against the gentle curve of its shoreline, and the lush rhododendrons that bloomed on the slopes of Haleakala. Now, a few months after my father's death, I was here once again to nurture my soul. I was hoping to find the dolphins and to experience some solace and rejuvenation.

☞☞☞

My father's death had been difficult. Twice over the holidays I had flown to Seattle — the first time, to spend what I knew would be our last Christmas together; the second, to sit with him in the hospital, and then, after his passing, to help my brother and sister take care of his affairs. On my first trip, I had gone to visit him in his condominium. He lived on Lake Washington, and on that particular day I could see the glistening slopes of Mt. Rainier from the large picture window in his living room. I understood why he had been reluctant to leave his home to move into an extended-care facility, and I knew that he was afraid. He was on dialysis and had recently taken a bad fall. As we sat looking at the view over the lake, we made small talk, mostly about his dog and two cats whom he was worried about. Then quietly, he said, "I'm not in control any more." "I know," I answered, "and it's okay. It's okay to let go."

On the second flight up from San Francisco, everywhere I looked I would see my father. I saw the back of his head as I stood in line at the ticket counter. I saw him sitting a few seats ahead of me on the plane, then again at the gate in the airport terminal. Praying that he would hang on until I got to the hospital, I could see him inside my head the way I remembered him in a photograph as a young man. He had on his army fatigues and looked strong and healthy and handsome. He was smiling.

When I did see him next he was in a coma in the hospital. The day before had been the last time he'd had full consciousness. With the help of the male nurse, I washed his chest, arms and hands, and shaved the stubble from his chin. I had never touched my father so intimately. Once he blinked open his eyes, but it was in startled pain as we tried to turn him. I sat with him all afternoon talking to him and reading, even singing. Dad lay quietly with his eyes closed.

He passed away quietly the next day, in the evening. My brother and I arrived shortly after he died. We had been eating dinner when we got the call from the hospital and had rushed through traffic. I sat quietly in the dark room, looked out at the beautiful night sky and the lights of the city and wept.

⌒⌒⌒

My first major experience of swimming with dolphins in the wild had been off the Big Island of Hawaii three years previously. Finding them had been an exercise in nonattachment, requiring a certain amount of trust and surrender. It hadn't been as simple as calling them up on the phone and arranging a date. It had been more subtle than that — more like missing and longing for a lover or beloved. At the time, it felt as though having a clear and sincere intention, along with a big leap of faith, helped in finding them. It also appeared as though I would get "messages," or a kind of acknowledgment, back from the dolphins in the form of sounds or pictures. On the first Hawaii trip I could hear them whistling in my head a week before I saw them. As I got closer to the dolphins, I was seeing vivid pictures of them in my head.

Before coming to Maui, I had "sent" a clear request expressing my desire to see the dolphins. I had spent my first day on the island with a dear friend, another dolphin lover, and towards evening we found ourselves at the beach gazing out at the splendor of a Hawaiian sunset surrounded by music and friends. As we watched the last rays disappear on the horizon, we decided to go to a concert at a house up in the hills. When we arrived, I was greeted by more familiar faces, smiling under the dim lights. I began to relax, breathing in the strong aroma of exotic flowers.

As we wandered onto the porch, I was introduced to someone new. He was a healer, he said. "What *kind* of work do you do?" I asked, wanting to get specific details. "Cranio-sacral" was the reply. I had experienced a session of cranio-sacral work once before and had enjoyed the gentle sensation of having my head slowly manipulated. "If you like, I'll show you," he offered. He didn't have to ask twice. I closed my eyes and slowly allowed myself to slip beneath his care. The tension of keeping it all together for the past few months through the final arrangements of my dad's passing slowly began to drop away. There was a sensation of weightlessness, and then I saw myself floating in a night sky of brilliant stars. Pure light pulsated from a million tiny orbs as I drifted through space. Then, through the stars, I saw water and a skyline off a rocky shore. I looked more intently and to the left saw the single splash of what I knew was a dolphin.

After awhile, I opened my eyes and, smiling, looked at my girlfriend. "I bet you the dolphins will be showing up soon," I said.

The next morning I was up early, driving with another friend to La Perouse, a bay surrounded by black lava rocks on the road south of Kihei. It's known by the locals as a place sometimes frequented by the dolphins. When we arrived, the sea was choppy. The wind was up too. Maybe this wasn't a good day. I started to walk across the lava anyway, my wetsuit in hand. When I'd walked some distance from the few cars parked in the lot, I gazed out across the water and focused my eyes to the left of the bay. There I saw a distinctive splash. That was all I needed to know. I squeezed into my suit and slipped over the edge of the rocks and into the dark water.

A large pod of spinner dolphins.

That morning I witnessed the largest pod of spinner dolphins I had ever seen. I watched, entranced, as they slowly danced their beautiful patterns. They swam in twos and threes, and then in huge groups, gliding gracefully across the bottom and then spiraling towards the surface. A few carried the little yellow leaves that they liked to play with. One dolphin playfully nudged the camera from the hands of a surprised swimmer. It was an amazing and breathtaking spectacle. As I swam with them, I couldn't help wonder if individual dolphins or pods ever communicated with each other on a spirit level over long distances, and if they were aware of some of the projects that humans engage in relative to them (my "project" being a book which had already taken hold in my mind). Somehow, with the dolphins, I had the distinct feeling that a certain amount of "intelligence" was sent back and forth through the ether.

Long, narrow rostrums are characteristic of the spinners.

The next day the dolphins were in the bay again. They were more rambunctious than on the first day; now they were roughhousing with each other and engaging in intense sexual play. They were closer to the shore, too, which made it a little less difficult to swim out to them. I laughed to myself — had they heard me moaning about my aching arms and legs the night before?

By the end of the morning, I felt complete. Shivering, I clambered over the sharp rocks to find the rest of my gear. I could feel every tender muscle in my body. It was time for a few days of "land" vacation. I felt that I had been given plenty to take in and integrate. Later, I heard that the dolphins had also taken a break. What perfect timing, I thought. I breathed deeply, taking in the gift

I'd been given. I felt a sense of renewal. I felt cleansed by the salt of the sea, the rare and transluscent beauty of the dolphins' dance and my remembrance of the starry heights that had figured so vividly in my imagination on my first evening. The grief and the ache in my heart over my father's loss remained, but the pain had lessened and what was left was held by a new sense of spaciousness. The future beckoned wide and open ahead of me. I knew I would remember these dolphins and their stellar performance for a long time to come.

CHAPTER 8

Play 101

The creation of something new is not accomplished by the intellect but by the play instinct acting from inner necessity. The creative mind plays with the objects it loves.

—Carl Jung

One of the central teachings I've received from the dolphins has been about play. Prior to being given that lesson, I was to be prepared for that experience by my cat Misha.

After he came into my life, there were crinkle toys everywhere — under the sofa and armoire and the living room chairs — and before long, a great stash under the stove. "Where's the crinkle toy, Misha?" I would ask, and he and I would embark on a patient search of the elusive spheres, usually ending in the kitchen. Crouched on all fours, my cat and I would peer intently into the dusty darkness beneath the stove, where the testy balls were hiding. I would recover them with the assistance of a long-handled broom, and our game would begin all over again — up the stairs, down the hall and into the closet. There was nowhere crinkle toys could not go.

In our early games with the crinkle toy, I would

throw, and Misha would fetch, carrying the ball back in his teeth and dropping it at my feet. However, we soon advanced to short games of toss where I would throw and Misha would bat the ball back with one paw or sometimes two. Soon we also discovered that we could roll the ball back and forth. The lazy cat's version was, of course, to roll the ball back and forth from a fully reclined position. Misha would dart one long paw out to expertly swat the ball just out of my reach, and then wait with one eye closed until I decided to return it. Hiding crinkle toys became another game. I would find them in my purse and in my slippers. Next to sleeping and eating, playing with crinkle toys seemed to be what Misha enjoyed most — it was his reason for being!

In hindsight, I was able to see how he prepared me for the play that followed with the dolphins. I've come to learn that advanced levels of play are about being fully present in an entirely unpredictable universe and opening to extraordinary magic.

∽∽∽

It was after I had finished my first lessons about play with Misha that Jon and I took our first group of people with us to the Caribbean to experience the bottlenose dolphins. All of us, including Jon and myself, the two marine biologists who we worked with, the trainers, the participants — and I need to include the dolphins — were very excited. For years, I had thought of bringing people to a situation where they could have the opportunity of having safe and extended contact with dolphins. They had become profound teachers for me. From my interactions with them, I intuited that many dolphins have an innate curiosity about people, especially groups of people. They are extremely social, with a highly evolved system of communication and sophisticated echolocation abilities. I was sure that humans presented the dolphins with as many questions as they posed for us. What would we be teaching each other if the two species literally went to school together? We would find out in our Dolphin Trainer for a Week program.

A young bottlenose dolphin blowing bubbles.

Part of the curriculum of our weeklong training consisted of snorkel-swim sessions with the dolphins. Wearing fins and snorkel gear, participants would swim in an area designated for the dolphins. "Remember to breathe," we told the participants. "Relax and let the dolphins come to you." I've learned that most dolphins are sensitive to the moods and emotions of swimmers. It was not until I'd mastered my own fear of this new situation that I began to have closer interactions with them. When I began to relax and enjoy myself — free-diving and playing in the water — the dolphins became interested.

"Another aquabat!" they seemed to say. When I entered the world of dolphinese, with my own versions of spinning, somersaulting and mimicking the dolphin kick, the dolphins recognized that they had a ready playmate.

I had noticed that the dolphins enjoyed playing with long pieces of seagrass that grew along the sandy bottom. During one of our swims, I dove down, picked a long strand and began dangling it from my fingers. Suddenly, a sleek form appeared at my side. I watched in amazement as the grass was deftly taken from my hand. I tried to contain my excitement, and I remembered Misha and how he loved the game of fetch. The next time a dolphin approached I tossed a piece of grass in his direction. The dolphin quickly caught it in his mouth and then, as he hovered there in the water, gently blew the grass back to me. I responded by returning the toss, and the game continued. What would Misha think?

On another occasion my eye caught the gleam of a white clam shell resting on the bottom. Diving down, I was able to grasp it in my hand and then slowly kick my way to the surface. When a dolphin came towards me, I threw the shell out in front of me. The dolphin caught it in his mouth, and another game of catch ensued. Suddenly, I lost my hold on the shell, and it slowly began to sink out of sight. Should I try to get it or will you, I thought. As if reading my mind, the dolphin gently spiraled down to retrieve the shell and then returned to hand it back to me. I was so surprised I began to laugh. I coughed and sputtered as my mask began to fill with water. Sensing an interruption to the game, the dolphin turned and with a flick of his tail was gone. I was thrilled by the interaction. I wondered how many other ways we could play together.

These and the many other games that the dolphins love to play have caused me to reflect on how some of our most intimate and satisfying moments with each other as humans are spent in play. It was apparent to me that for members of the animal kingdom as well, the area of play is a place of subtle and intimate bonding that goes far beyond our thought processes and mental understanding — it's a place of enchantment and deep joy.

CHAPTER 9

Joyful Interludes

To the dolphin alone, nature has given that which the best philosophers seek: Friendship for no advantage. Though it has no need of help from any man, it is a genial friend to all and has helped mankind.

—Plutarch

A s our weeklong trips to the Caribbean evolved, and I was able to spend more time with the bottlenose, they began to reveal themselves as having unique personalities and gifts, similar in many ways to humans. As a species they seem especially vivacious and extroverted, but some of them are actually shy and more inward. Many of the dolphins I met were highly skilled and talented in learning new behaviors. A few were not that interested. Different dolphins had games that they particularly liked. They had buddies, they had disagreements with one another, and sometimes they argued dolphin-fashion. They had good days and not-so-good days. They had moods; some of them were hard workers, others were more lazy. I remember one dolphin being especially skittish and high-strung. She was jealous of a young dolphin named Tela (Tay'lah) who easily mastered new skills in the training sessions. She would deliberately race by and whack Tela with her fluke.

Most of the dolphins recognized their trainers and the staff; they even seemed to remember and bond with individual participants during our sessions. Several dolphins made particularly strong impressions on me.

↦↦↦

Tela is a female bottlenose dolphin. At the time of our first meeting she was almost two years old. She had an impressive lineage. Her mother, Rita, is the alpha female of the pod of dolphins with whom Jon and I had begun to spend time. Tela's father is Copan — "the King." I had met Copan on my first trip to the island. Not only was he by far the largest dolphin of the pod, over nine feet long and weighing more than 600 pounds; Copan also had a large, compelling presence. He ruled. He commanded respect and the other dolphins deferred to him. He was revered and genuinely loved by his human trainers. He was the stuff of myth and legend in his own time. Ultimately, Copan was transported, along with several of his favorite females, to another facility. At last he could have the amount of space he deserved. "We all cried when Copan left," one of the trainers said.

Given that the gestation period for dolphins is about 12 months, Tela probably was conceived a few months before my first encounter with Copan. When I met Tela a few years later, she was a wild child, full of light and flirtatiousness, humor and intelligence. She radiated joy and mischief. I wrote poetry about her and smiled whenever I thought about her. I remember our feeding sessions together, which became a time to show off and engage in outrageous play. She seemed to know all the behaviors and then some, even without formal training. Eating a fish was an excuse to flip it high, jump, fall down, spin and catch it all at the same time. Tela was a giggler. Her beautiful lined gray markings ran from the end of her rostrum over her melon (or forehead), to the blowhole on the top of her head. Today she has a distinctive rake mark, consisting of three parallel lines, on the lower side of her dorsal fin.

On one visit I discovered that Tela had learned to dance — a girl after my own heart. Dancing rates high on my list of favorite things to do. Her trainer, a young island man, wore a constant wide grin across his face. His teeth and the whites of his large brown eyes shone brightly as he joked and kept up a continuous stream of banter and tall tales. He waved his arms and legs enthusiastically to emphasize his stories. It's no wonder that Tela happily wiggled about as her trainer sang and rocked his lithe torso in front of her.

One of Tela's playmates was a male dolphin about the same age, whose name was Dexter. Dexter was another charmer. He would give me his undivided attention during a training session and then be off and running, performing behaviors to perfection before I had time to even begin a hand signal. It's as if he clearly understood my intention or the meaning behind my words without further need for interpretation.

Another one of the youngsters in the pod was Li'l Bill. During the first year of his life Li'l Bill's constant companion was Maury, an inquisitive and outgoing female dolphin just a few months older than himself. While Maury was social and curious, Li'l Bill was quiet and shy. He was always in her shadow — smaller, slightly darker in color, more withdrawn. His mother,

Samantha, whose underbelly showed the long scars of an old shark wound, died suddenly after our first trip. I was deeply saddened to hear of her passing. I remembered Samantha as being a particularly loving and patient dolphin. The trainers wondered how Li'l Bill would react to his mother's death. Would he become even more reclusive? Would he separate himself from the pod's social structure of interdependance and nurturing? He surprised everyone. He adapted to his new independence with firm resolve. In one particular encounter session, Li'l Bill waited patiently until he had the full attention of each participant, who would master each hand signal perfectly, before doing the behavior. He had become the teacher! It was apparent to all of us that we were having a lesson on being fully present and giving clear and direct communication. I was told that in another session Li'l Bill had exhibited a new behavior. He had given a long heart-to-heart hug to one of the participants as he stood vertically in the shallow water.

At the end of my second trip to the islands, the day arrived when we were scheduled to have our Farewell Encounter with the dolphins. Most of the participants stayed near the shore to work in smaller groups. Three of us returnees were given a chance to work more intimately with some of the other dolphins. At first I challenged myself to work with two of the older males to learn the "rocket ride." To perform a rocket ride two dolphins would put their rostrums on the soles of a person's feet, pushing them through the water at high speed. When ready, they would dive to the bottom, the dolphins still pushing, then turn upwards to break the surface and fly into the air.

On this occasion, the two males, Estaban and Paya,

were all business. They rose, straight and imposing, out in front of me — nothing soft about them. It proved to be more than I wanted to handle, trying to coordinate both of them and myself. I decided that I might feel more comfortable working with one dolphin, preferably a female! I asked a trainer if I could work with Gracie, Maury's mom, who was nearby. As I approached, she snuggled up to me; again and again we practiced as she pushed me away from the dock. At one point we came to a stop out in the middle of the water. When I turned to find her, she planted a big kiss on my cheek with her rostrum and then gracefully turned over to have her belly rubbed. *Let's take a nice break and enjoy.* I couldn't have agreed more!

Kissed by a dolphin.

As the morning came to an end, we all strolled down the catwalk to where a boat would take us across the lagoon. There was an air of sadness as we all anticipated our leave-taking. As we stood holding our wet clothes and gear, we saw the dorsal fins of three dolphins as they crossed the water to come toward us. As they approached, we were able to identify them. There was Tela holding a perfect yellow leaf in her mouth. Next to her was Dexter carrying a long piece of seagrass, and behind him was Li'l Bill. It was hard to believe! All of us went down on our stomachs on the wooden planking, our heads leaning down toward the water. For yet another ten minutes we stayed and played. Good-bye sweet Tela and Dexter. Good-bye Li'l Bill. Thank you.

I thought that would be my last time with the dolphins that week, but the next morning I realized I had finished my packing, eaten breakfast and retrieved my passport, and I still had an hour before we were scheduled for pickup by the bus that would take us to the airport. I hailed the taxi-boat and headed out to the key. I shouldn't have been surprised to find several other people with the same idea. I asked if I could help feed Gracie. I was handed a bucket of fish. "She's all yours," was the reply. As I went to toss Gracie a fish she rose out of the water and gave me a big kiss. Throughout the feeding I kept getting kisses. Hello and thank you. Thank you, thank you. My lips were bruised, but I couldn't have been happier.

CHAPTER 10

Captive Versus Free

The worst sin toward our fellow creatures is not to hate them, but to be indifferent to them; that's the essence of inhumanity.

—George Bernard Shaw

The question about dolphins being in captivity is one that has provoked much controversy. What I experienced on the island is perhaps an example of dolphins living in a situation where both dolphins and people have the potential to benefit. What impressed me the most was the strong bond of affinity that existed between the trainers and the dolphins. The trainers here are mostly young locals, and each trainer is responsible for one dolphin. It's a relationship based on trust, respect and patience, and it's a huge commitment in time and energy.

The dolphins live in an enclosed portion of a natural coastal lagoon adjacent to a tiny tropical islet. They receive a daily diet of fish, each dolphin receiving a specific ratio of herring and capelin, depending on their specific needs. Obviously, the availability of fish is a primary incentive for taming and training dolphins, but I wondered if there were other motivating factors that kept the dolphins from leaving their island home. I asked one of the marine biologists about whether or not any of the dolphins ever left the pod or lagoon. "Of course," he said. "We keep them contained, but there are plenty of opportunities for them to escape if that's what they want to do." He went on to explain that in all the years he had been with this pod he remembered only one dolphin ever having left and not returning. That dolphin had not been able to form social bonds within the group and had finally decided to look for greener pastures elsewhere.

It's the juveniles who most often tend to run off, leaving for days at a time. "They need to sow their oats," he went on. He told the story of one of the younger males swimming after a passing pod of wild dolphins. The marine biologist followed in his boat, worried that he was going to lose his dolphin; he continued following the pod way past the end of the island. Suddenly a call came in from the owner of one of the local bars on the West End. "We've got a dolphin at the end of the pier here. He's being awfully friendly and we wondered if it was one of yours," said the caller. The boat was quickly turned around and raced back to the pier, but the dolphin had disappeared. Disappointed, the biologist returned to the key only to find the happy wanderer waiting outside the gate to be let back in.

Another story we heard was about the huge hurricane that hit the islands several years ago. The damage to the two small keys in the bay had been severe. During the storm the center had decided to set the dolphins free since many of the docks and structures by the water were being destroyed. In the weeks that followed, the resort began receiving phone calls from people all over the island. The story was the same. The dolphins were showing up in harbors and alongside the wharves looking for home.

Personally, I've never been able to bring myself to visit captive marine life in an unnatural setting. As a human I find it difficult enough to swim in chlorinated pools, and I can't imagine dolphins or other captive marine animals living in a pool with concrete walls that do not allow them to fully utilize their sonar. I also know that dolphins separated from their offspring, or babies from their mothers, would probably die — most often from depression. Single dolphins do not tend to do well. Their chances of survival increase significantly when they have companions or are part of a larger pod. They're not so different from people, I thought.

I know that there are compassionate trainers everywhere, and I'm sure that there are some who are not. I am also aware that marine mammals, just like some of their human counterparts, are exploited purely for financial gain. On the island the setting is much different from what I suspected was true of some marine parks. The dolphins live in a natural habitat in a lagoon. Food and protection are offered in lieu of their freedom. One could also argue that a safe harbor for a pod of dolphins is no small thing — in the wild, many baby dolphins do not survive because of shark attacks. I listened to the story from the trainers about dolphins being out on a dive and suddenly returning to the safety of their pen because they sensed danger.

I am also aware of some of the rehabilitation programs that help to reacclimate dolphins to the sea. Dolphins need to relearn how to catch fish and fend for themselves. I thought about my cat Misha and how he would manage without me. Misha is so gentle that it was hard to picture him catching, much less killing, mice or birds for a meal. He has evolved into a different kind of cat from his ancestors, prefering the company of a human to roaming untamed savannas.

Dolphin responding to a hand signal.

The time normally spent on avoiding predators and hunting for food was being used by tamed animals to learn new skills and behaviors. What would I do, I wondered, if I weren't constantly "hunting" for *my* next meal? The two scenarios, I know, are not analogous; however, there are some interesting parallels. What interested me more was the effect that dolphins (as well as other animals) and humans can have on each other as they coexist in a kind of community. I could see the beginnings of this process in the attitude of the trainers. These young boys were exceptionally friendly, sharing an easy comraderie and playful humor that was contagious. Through their jobs as dolphin trainers they had become an intimate part of a more integrated way of life involving the water, the weather, the seasons and the animal kingdom.

I was also intrigued by the effect the dolphins had on the participants in our week-long programs, and vice versa. It's impossible to remain indifferent to the beaming smile of a dolphin. Universally, dolphins are known to evoke deep emotional responses from their human friends. I'd seen people in our groups arrive emotionally shut down and on the defensive. As the week progressed, they would open up in astonishing ways simply by having been in the presence of a dolphin. I remember two young women who had traveled all the way from England to be with the dolphins. One afternoon they were so moved by the unbelievable scene of humans playfully interacting with these huge mammals that they sat in the shallow water weeping, overcome by the rare beauty of such an extraordinary event.

I observed a new kind of language emerge as the two species interacted with one another — a kind of soul language — a language of laughter and excited shouts punctuated by high whistling and splashing. And just as we humans shared our adventures together during our daily meals, I suspect that the pod communicated their experiences with one another. I remember lifting my head up out of the water during one of our final snorkel swims. I could see at least five different pairs of dolphins and humans playing catch with the seaweed. When did everybody learn that, I wondered.

In another session we were enjoying the dorsal tow rides. As it came time for one of the men to take his ride he was unceremoniously dumped into the water by the dolphin. Later another dolphin did the same thing! He was left bobbing along by himself. During a sharing we had later in the evening, he admitted that he'd learned a powerful lesson that day. He hadn't seen the dolphins as much more than animals that needed to be controlled. The dolphins in turn had collectively agreed to ignore him.

A swimmer and a dolphin in silhouette.

Of all the dolphins, it is the young dolphins who are particularly fascinating. From the time they are born, they are interacting with both humans and dolphins, and not knowing anything different, these youngsters exude a fearless curiosity that is at once compelling and unnerving in its innocence. In one of our swims I remember suddenly seeing a small white face pressed up against my mask. Its tiny white flippers stuck out to either side like the little wings of a watery angel. I remember being puzzled, not able to recognize the newcomer. Later I learned that this was the new baby of the bunch. The youngsters are amazingly receptive and eager to learn. I'm sure they are affected and influenced by the humans interacting with them, learning with fierce rapidity all that our spirits have to teach them. And as with every new generation of humans, it is the new dolphins who appear to be leading the way and evolving into newer levels of awareness.

In an ideal world, or perhaps in some future ideal society, dolphins and humans will live next to one another on the land and in the sea in harmony and without containment. They will play with one another and share knowledge and information. They will live without fear and in peace. They will have some common language, perhaps in pictures, whereby they will communicate.

Until that time I am convinced that all of us — humans, animals, the plant kingdom and the changing elements of wind, water, fire and earth — are more complete and whole if we accept the fact that we are all interrelated and that we can learn from and help one another. Humankind does not have the monopoly on truth and could benefit from the wisdom of the other natural kingdoms. Hopefully, as the different species learn to listen more carefully and deeply to each other, we can develop true alliances that are worthy of respect.

CHAPTER 11

The Starry Realms

The dolphin signifies higher consciousness, the connection
with the stars, especially Sirius.
                    —Celia Fenn, from *A Message for Lightworkers*

During one of our evenings on a weeklong dolphin
trip, Jon invited the group to participate in a meditation, or guided imagery process, with the dolphins as
our focus. "Just pretend or imagine for a short time that
the dolphins have a spirit and a consciousness that you
are able to connect to and communicate with," Jon
began. With our eyes closed, quietly sitting, he asked us
to invite the energies of the dolphins into our space.
"Notice in your mind's eye if you see them," he continued, "or if you hear or feel anything unusual. Notice if
you see any particular dolphins that you recognize, and
see if you are able to make eye contact. Ask them if they
have anything they are wanting to teach you or show
you. See if they are wanting to take you on a journey."

In my mind I saw the large shape of a dolphin appear. I knew immediately that it was Copan, the alpha
male I had first met on the island. He came to stand
behind me near my right shoulder. I could feel his

strength and the wisdom of his spirit as a leader quietly flowing into me. Then, behind me to my left, I could feel a second presence. This energy was lighter and more playful. It sparkled with flashes of humor and graceful femininity. I knew right away that it was the sweet being of Tela.

"Come," said Copan. I grasped his dorsal fin in my hands, and we began our flight through a shower of blinking stars. The brilliance and beauty of the scene was indescribable. The spacious- ness and serenity of this endless sea of lights gave me a feeling of utter freedom and complete peace. I was traveling through time and space on a journey whose destination was unknown to me. I was aware only of the joy and magic of our flight.

Author's depiction of her meditation.

As we shot threw the heavens, Tela flew a ring of perfect spirals round and round my body so that I felt like a caterpillar encased in a gossamer cocoon. Suddenly, we burst through yet another band of shimmering star systems, and I saw myself standing as a body of light. All around me the dolphins began weaving a complex coat of horizontal figure eights — miniature infinity symbols in shining gold. I felt myself at the center of a vortex of pulsating energies.

From far away I heard Jon's voice telling us that when we were ready we could think about starting to make the journey back home. I held on to Copan, and we began to retrace our flight. "Is there anything else you'd like to ask your companions before you thank them and say good-bye?" said Jon.

I could feel myself coming back to earth. Yes, I thought. "How will I know how to do the right thing?" I asked.

"Unconditional love" was the answer. "Unconditional love."

I thanked my guides and slowly opened my eyes, returning to the room.

↬↬↬

I have been curious about what seems to be a connection between the dolphins and the stars as they've appeared in my meditations, and on several occasions I have seen and heard references to there being a link between the star system Sirius and the dolphins. A few years back I went to a talk by a former NASA employee. He showed video footage of unidentified flying objects as they made their curious appearances in the night sky. The speaker described the shape of some of these flying objects and then related a myth about the Dogon tribe in Africa. The myth tells of visitors from the

star system Sirius. Drawings of this star system and of
half-aquatic beings have been found on the walls of some
of the caves in the area of the Dogon people, along with
round stone artifacts whose shape resembled the unidentified craft we had seen on the footage.

How much truth there is to these stories is hard to

⊱ *Coming Home* ⊰

Good-byes are often hard to do and, in the best
of circumstances, transitioning back to our daily lives
after a vacation or attending an inspiring seminar can
be challenging. For many people, saying good-bye
to the dolphins and going back to life without them
was that much more difficult. On one occasion, a
large group of us were on the same flight back to San
Francisco. As we approached Customs and Passport
Control in the tall glass-and-steel rooms of the airport
terminal, I could see by the expressions on the faces
of our group a new kind of softness and vulnerability.
I heard someone say, "I feel like we're coming back
from the planet Venus. The planet we just arrived on
could be Mars."

Because of these experiences, Jon and I had
devised a little game, or visual exercise, to help people more easily re-enter everyday reality. During one
of our final evenings on our trips, we would invite participants to join us in a guided visualization. After
people had relaxed and found a comfortable sitting
position with their eyes closed, Jon would begin to

know, but I do find it fascinating that tales like those of the Dogon tribe exist. What seems important to me are the deeper awakenings that can be stirred in the human heart and soul through communion with a higher form of spirit — spirit embodied in more evolved teachers, other people, kindred animals — even waterfalls, trees or

speak. "Pretend that you have the possibility of inviting some of the dolphins home," he would say. "In your mind's eye see your home — where you live — and ask some of the dolphins, or perhaps the whole pod, if they would like to visit. See them entering through the front door of your house and swimming through the rooms. See what it feels like to have them in your space. Invite whatever qualities you feel they embody to permeate your life so that you can continue to experience them even after leaving the island." In silence, we allowed people to bring the dolphins home.

After returning from one of our later trips with the dolphins, I was surprised and delighted to hear about the results of this practice. Before leaving for the week, a client of mine who had been visiting from out of town — a dolphin lover and a particularly sensitive artist — needed a place to stay for a few days. I had offered her the use of my home while I was gone. When I got back, she looked at me curiously and said, "While you were away, it felt like your room was filled up with dolphins!" I had to laugh. "Yes," I said, "I sent them all home."

mountains. What is apparent to me is that we can form alliances with beings that can assist us on an energetic level in understanding ourselves and in making choices. We can, through directing our attention, appreciate their positive qualities — their strength, beauty and wisdom.

In this way we create allies, and our alliances can change. We can have guides for certain projects or for help in creating particular outcomes. Allies can help us tune in to guidance, as well as in receiving nurturing, healing and inspiration. It can be a life-transforming experience to be in the presence of a dolphin in its native environment, and it can be equally uplifting to connect with the spirit of the dolphin through our feelings and focused intentions — sometimes even more powerful, since we can be less distracted by their physical form.

As Jon and I were returning from our first visit to the Caribbean, we were as high as kites. Jon suggested that I use this feeling of ecstasy to create something I really wanted in my life at that moment; this was the time to put into practice the laws of manifestation. From such an expanded state, dream a bigger picture, he said, without a lot of mental worrying about how it might come about, or doubting that it could happen.

Just before we had left on that trip, I learned that I would be needing to move from the house where I had been living for three years. "What timing!" I groaned; "I'm leaving on a trip and won't have time to go house-hunting." With Jon's assistance, I began to "paint" a very clear picture of exactly what I was looking for in a new home.

"I need light," I said, "and I would like a private deck

with sun, white carpeting and high cathedral ceilings, a garage, a compassionate landlord who likes cats, and someone to share the expenses with me." In my mind I could feel myself in this new space with the carpet under my feet and the light coming in from the back. I allowed myself to feel how happy and secure I would be.

"Okay," said Jon, "Imagine that you can have that. Now if you could have ice cream on top of your cake, what else would you ask for?"

It took me only a second to reply, as I thought about the dolphins. "Water," I said. "I would like to be near water or a place to swim."

Within three weeks of returning I had found my new house, along with the perfect housemate. Anything I was missing in the way of furniture or appliances she had. There was also a small private pool two blocks away, and it was two minutes to the water where a beautiful bike path followed along San Francisco Bay. Now, almost every day, I start my mornings with a walk by the water. I watch the seagulls and the herons, and I can see Mt. Tamalpais and the Richmond/San Rafael Bridge in the distance.

CHAPTER 12

The Emerging Feminine

The Divine Mother is the one who gives us everything in the world. To Her you should offer your love, your devotion, your worship, your work. You should offer everything to Her and feel grateful that you are given the chance to live in this world, because it is Her grace.

—Shri Shastriji, Indian poet

My first significant meeting with a dolphin at the marine biology center had been with Copan. He was a wise, strong and capable, masculine presence. He seemed to know exactly what he was doing. His spirit seemed to radiate an awareness of the importance of what his task was as an intermediary between species. My first interaction with Copan had been one of the deciding factors that had led me to bringing people to the island for an experience with dolphins.

A few years later, when I learned that Copan had been transferred to another facility, I was disappointed, knowing that I would miss his presence. Immediately, as if in response to any second thoughts I might be having, I "knew" that there would be many new dolphins being born, and that we would all have a lot to learn from one

another. Later, I understood that the balance of power
had shifted from a masculine to a feminine energy in the
pod.

When Copan left, the alpha in the pod became Rita,
and as Copan had been king, she was queen. There were
two other males who vied for the position of dominant
authority, but Rita was older and the rest of the pod gave
her their respect. She was another very large dolphin. No
one else ate their fish before Rita. She had first choice in
all matters of importance. The trainers were always careful to include her in any major activities and changes. As
one of the leaders of our human pod, I realized that I
would have to introduce myself to her. I got to know her
during a feeding session. She was strong and demanding,
but under her tough facade I sensed a genuine sweetness.
Her daugher Tela was a true princess — effusive,
excitable and royal — a radiant light-being.

Rita and Tela.

During our training programs, most of the pod that
we interact with are the females and their offspring, and

the energy has been distinctively feminine. All of us have observed the qualities of nurturing, caring and training of the young. I've also noticed a distinct softness and vulnerability to our training sessions. I've sensed that this emphasis on the feminine and on feelings is appropriate now, to these groups and to our times. Learn and know through your feelings. Listen to your heart. What is your heart telling you? Communicate and create from your heart.

As a society it seems we have become polarized towards the head, often living from fear and manipulation. Our focus and values have been on the thinking processes — the "smartest-will-win" kind of attitude. The way of the heart has been dismissed as not being an effective means to achieving power. It is too unpredictable and often illogical. It often seems to go in circles as opposed to straight lines. It's confusing. As a result, we seem to have lost some of our trust and knowledge in what is truly good and of value. The feminine principle concerns itself with interrelationships and interdependence. It knows that healthy connections are necessary for our survival. Every pod of dolphins knows that. The health of each individual dolphin is important if the pod is going to protect itself against predators and if it is going to consolidate its efforts for catching food. The caring of young dolphins is often shared between several mothers or aunties.

Our human society, in contrast, seems blind to these values. Childcare, our schools and the environment always seem to end up at the bottom of the list in our national priorities. A grossly unfair distribution of food and resources leaves the major portion with a tiny elite. Our money and attention are directed toward achieving "security" through our destructive capablities. What good will an overabundance of things do if we no longer have the heart to appreciate them, the compassion to include others into our lives, or the common sense to stop destroying ourselves through arms races and environmental degradation?

My heart breaks when I hear about the low-frequency sonar testing that is being done by the navy in our oceans, and how dolphins and whales are found beached with their eardrums ruptured because of internal hemorrhagging caused by extreme sound levels. When I hear about farmed salmon being fed massive amounts of antibiotics and food from artificial sources that cause them to die of disease and a weakened immune system, my heart breaks again. When I hear about the reefs of our oceans that were once rich ecosystems of food and life being stripped and laid bare, I know something is deeply wrong. To me these are signs that our connectedness and appreciation of other life forms is being lost. It's difficult to imagine a world without dolphins and whales swimming in our seas.

My guidance tells me that the feminine spirit and intuition needs to be reawakened in all of us. Only when we tap back into the wisdom of our hearts will balance be restored; only then will we be able to heal our planet and ourselves.

CHAPTER 13

Beyond the Veil

*And when the earth shall claim your limbs, then shall you
truly dance.*

—Kahlil Gibran, *The Prophet*

Sometimes the threads of our lives weave themselves
into patterns altogether strange and mysterious. When
the design of a life is suddenly interrupted or ended, a
shifting occurs, and in that place of opening and vulnerability miraculous breakthoughs in understanding are
possible.

Looking through my e-mail one day, I saw the words
"In Memoriam" without at first having them register.
The message was about a good friend — a fellow traveler,
a seeker of truth and dolphin lover who had just passed
away. He and I had shared much in the way of play and
our love of the water and the dolphins. I knew he had
been in Maui finishing work on another music album,
and I had heard that he had met a young woman while
swimming with the wild spinners in Hawaii. They had
fallen in love and were very happy. Then, quite suddenly,
he had taken a fall and passed into unconsciousness. He
died a few days later, surrounded by friends.

I sat at my desk in my pajamas with my coffee. He had died the day before. As I stared through the glass of the sliding doors to the greenery and blossoming jasmine on the deck, my consciousness seemed to be operating on several levels at once. A part of me was attentive to a project I was finishing for a client. Another part was reaching out and up through time and space and asking *why.* And, *are you all right?* The last time we had spoken had been difficult. We had needed to work through some misunderstandings, and since then I knew I had created some distance in our relationship. Now I felt a slight sensation of vertigo. All those questions — "would have, could have, should have" — hung at the edges of my mind. I took a deep breath, and then in my mind I could see him smiling. *I feel great. You should see me now! Don't worry — have courage. All is forgiven and I love you.* I felt like I had just been bathed in a shower of warm light.

For the next three hours I worked on my project. I also watched as the beauty of the sunlight glinted off the bright green foliage outside. I could hear the sweet sounds of the birds in the trees. I saw the soft rippling of the wind as it stirred the delicate flowers on the vines. I needn't have worried about him. It felt as though he was surrounded by a host of compassionate beings tenderly assisting him in his passage. I thought about the New Testament story of Mary Magdalene arriving at the tomb to find her Beloved ressurrected into a being of Light. I remembered my cat when she died, gently becoming one with everything, her spirit expanding into all that is. I felt at peace.

At a memorial gathering a few weeks later I was struck by how many people had strong and intimate experiences of this man after he had passed on. He seemed to show up simultaneously to many different people. Strange to say, yet true, in some mysterious way I feel closer to those I love after they have passed on than when they were in the body.

At another memorial event for this man, a group of friends and family met to walk together to the ocean. We spent the sunset hours sharing stories about our friend's life and listening to some of his songs that we played on a portable boombox. On the way back four of us stopped spontaneously and hugged one another as we stood together on the road. At that moment the radio seemingly turned itself on and started blaring out the song "Lean on Me." We burst into laughter and began to sing along in high voices.

As the words of that song suggest, I felt that, in this moment of sorrow, we were being invited to "lean" on our friend who had recently passed on, even though he was no longer in his physical body. And we were also being reminded to remember that there is, indeed, "always tomorrow."

This is good, I thought, as are laughter and the gift of friendship. And, remembering to breathe... breathe... breathe...

CHAPTER 14

Orca Sightings

The Mowachaht/Muchalaht people believe [the killer whale Luna] embodies the spirit of their late Chief who had told several people he wanted to return as a killer whale and stop the fish farming in Nootka Sound. The whale showed up alone in the sound the same week the chief died in 2001.
—Canada's Department of Fisheries and Oceans (DFO), Press Release, June 2004

It was inevitable that my fascination with dolphins would come to include other marine mammals. As it happened, my attention was drawn to the orcas — some of the most distinctively marked and charismatic mammals in the ocean. I was happy to learn that *orcinus orca* is considered part of the Delphinidae family. Although often referred to as killer whales, they are in fact the largest of all the dolphins. They are known primarily for their intelligence and fearless hunting skills, and they live in small, tight-knit, lifelong pods, each ruled by a matriarchal female.

As I delved into orca lore and history, I became especially interested in those living in the Northwest and in the waters off British Columbia. Sometimes accounts of individual orcas would make headline news. One of these was about an orca called Luna. Luna was a young juvenile who had apparently been separated from his pod and was now living by himself in Nootka Sound, a bay halfway up the western side of Vancouver Island. Because he was lonely, he started frequenting the docks of small coastal towns, introducing himself to boaters as they passed by. Soon he was considered a nuisance by local authorities. Word of his predicament spread, and I heard that an effort was being made by several animal communicators to try and reunite the wayward whale back with his family. Curious, I decided to try out my own long-distance animal communication skills by checking in for myself to see what I could pick up. As I put my attention on Luna, I received a clear image right away of a very active and sociable young orca. When I tried to put a "signal" out to his family, I got nothing, but I kept trying on successive days. Finally, I received a faint impression from what I felt to be a pod somewhere much farther south. Then I heard, *We hear you and we're busy right now. We're fishing. We're not too worried about Luna. He has his own path to follow.* I was surprised by what seemed a lack of concern. Since then, I've learned that, like other dolphins, most orcas put fishing at the top of their list of priorities.

As my interest in the orcas grew, so did my desire to see them up-close. Opportunity presented itself in the form of a kayaking company that needed some Web design work done. I agreed to barter for a kayaking trip. We would be paddling in Johnstone Strait off Vancouver Island, the favorite hunting grounds for the northern resident community of orcas in the summer. That's when the orcas are attracted by the salmon runs in the straits.

⌐⌐⌐

The orcas of British Columbia and northern Washington State are some of the most researched orcas in the world. Scientists have divided them into three distinctive populations: *resident communities,* of which there is a northern and southern community; *transients;* and *offshore* orcas. The *resident communities* live on fish, primarily salmon, and travel in very specific areas. The northern community covers an area of approximately 500 miles, including the area surrounding northern Vancouver Island and extending as far north as southern Alaska, and consists of approximately seventeen pods of about 160 orcas. The southern community ranges around the southern half of Vancouver Island and into the San Juan Islands and Puget Sound. This group is made up of three pods of approximately 80 orcas. The *transients* seem to travel in both of these two regions and prey almost exclusively on smaller marine mammals, including seals, dolphins and other small whales. The *offshores* appear to live in the open ocean, feeding on fish. The three groups rarely intermingle, and although their paths sometimes cross, they appear to avoid communication and contact with one another.

The northern and southern resident communities have been extensively studied. Many orcas have been individually photographed and identified by their dorsal fins, which are distinctively shaped and have unique markings, nicks or scratches on them. Scientists have devised a naming system whereby each pod is given an identifying letter, such as "B," and each orca within a pod has a number.

One of the more fascinating aspects of orcas, like other dolphins, is their communication with one another and their demonstration of sophisticated patterns of vocalization. Along with echolocation sonar "clicks," each pod seems to have a dialect unique to that pod, along with individual signature whistles. Related pods share overlapping calls and whistling sequences.

~~~

The kayak trip was to begin at Telegraph Cove,. a 20-minute car ride from the small town of Port McNeill on the north end of Vancouver Island. I decided to fly to Seattle from San Francisco and then drive up and across the Canadian border and take the B.C. ferry over to Nanaimo. From there I took the scenic highway up the eastern coastline of Vancouver Island, passing long expanses of beach covered with mussel shells, clumps of bull kelp and the bleached remains of scattered driftwood. Clusters of purple lupine hugged the roadside, along with wild white and yellow daisies and Indian paintbrush. Later the road passed through more mountainous terrain, and when I stopped to refuel, the young Canadian attendant gave me an updated weather report.

"Wind's been picking up," he said. "Might be a bit choppy out in the strait for paddling, eh?" He clipped his words and pronounced the word "out" with the long "o" sound typical of Canadians. As I got back into my car, I sent an inner message to the orcas. How about some calm paddling for a novice? In my mind I saw the tall dorsal fins of three orcas traveling through a wide passage of gray water.

At our put-in site the following morning, I gathered

with the other ten participants and our three guides to load up all our gear. Soon our kayaks made a festive display of color — light blue and green, yellow and magenta — as we headed out into the strait. The skies were clear, and the water like glass. Prior to the trip, I had few expectations other than the desire to see orcas. What I had not expected was the variety of wildlife we were to encounter. I would come away from the trip with a new appreciation for how we humans are just one small piece of a much bigger and complex puzzle.

Our trip began through a maze of tiny islets and floating islands of kelp. After a breezy stop for lunch at a small beach, I gradually fell into the rhythm of paddling and was soon lulled by the steady rise and fall of water droplets as they fell from our paddles. Suddenly, in the distance I saw a series of graceful scalloped arcs cut across the water. The strings of little silver jumps appeared everywhere. They were Pacific white-sided dolphins. Gregarious and acrobatic, they're smaller than the bottlenose dolphins and are identified by a white streak or

Pacific white-sided dolphins.

stripe along their sides. Apparently, it was unusual to see so many of them. For several years one of the dolphin's favorite foods, the pilchard — a small herringlike fish that swims in huge schools off the coast — had disappeared. Now they were back in swarms, luring the dolphins into the inland passages. Later that day we also saw Dall's porpoises, their tiny triangular dorsal fins appearing and reappearing at regular intervals. Toward late afternoon we pulled our kayaks up onto a beach popular as a campsite for kayakers. The year before, the guides had been fortunate in witnessing an unusual event here: several orcas had moved into the shallow waters to rub their pectoral fins along the gravelly bottom.

As dusk began to fall, we could see the silhouettes of fishing trawlers and hear the hum of the salmon purse seiners hauling in their huge nets. It occurred to me, as I contemplated the scene, that the main attraction in the strait, despite the abundance of different wildlife species here, was the salmon. Hunted by humans and orcas alike, they are a dynamic force that determines both the lifestyle and politics of several intricately woven life forms.

Northern resident orca breaching near Vancouver Island.

The moon that evening was full, sending a faint shimmer across the wave tops. A family of busy sea otters frolicked in the kelp beds right off shore. As I sat in a reverie, there was an abrupt shout from one of the guides — "Shhh... listen." Our attention was directed to a point off to our right. There, jumping completely out of the water in front of one of the fishing boats, was the graceful form of an orca. Its black and white torso gleamed in the pale light. We watched in amazement as several dark dorsal fins moved past us to an outcropping of rocks on the opposite side of the beach. "They're moving west!"

Orca spyhopping.

Scrambling over one another, we clambered up onto the slippery rocks for a better look. As we stood there in the twilight, an orca spyhopped (lifted its head vertically) straight up out of the water! We shrieked wildly, waving our hands over our heads — "We're over here!" Again, the orca rose out of the water. We shrieked again, cheering and waving. For a third time, the orca came straight up out of the water, curious about all the commotion. Like a group of wild banshees we continued to jump up and down, brandishing our arms in the moonlight.

The next morning I awoke early. I struggled out of my sleeping bag and, not wanting to miss a moment, stumbled down to the beach. The rays from the early morning sun were blinding, bouncing off a glassy sea. Fir-lined islands floated on a silver calm still wreathed in quiet mists. I sipped a hot cup of coffee, enjoying the simple serenity and beauty of the hour.

The plan for the day was to continue down the strait to a sheltered reserve named Robson Bight (Michael Biggs) Ecological Reserve. Established in 1982 as a marine sanctuary for killer whales, its purpose is to provide a safe habitat for orcas where they are protected from harassment by humans. The area is off-limits to boaters, but often it is possible to see orcas passing by on their way into or out of the reserve. As it turned out (and as reported on the local "whale channel"), the orcas were busy fishing and socializing that day, so our paths did not intersect.

The fluke of a humpback whale.

On our third morning, after crossing the two nautical miles of the strait, a message came in over the radio. A humpback whale had been sited and was heading our way. Staying close to shore, we patiently decided to wait for its arrival. A loud booming sound soon caught our attention, and our eyes were riveted by the sight of an enormous spout exploding in the distance. Gradually, the giant mammal moved towards us, its great fluke creating a magnificent display of waterworks. Slowly it passed by us, on the other side of the channel, its long body rising and falling below the waves.

The next day, I wondered if we would be seeing more orcas. My instincts told me that they were down in Johnstone Strait, and we were headed in the opposite direction. I bet we'll see them on the last day, I thought to myself. The day was spent kayaking through a chain of small islands and waterways. We passed brown bears foraging among the rocks onshore, eagles in their airy nests, and the bobbing heads of curious seals. We also saw more of the graceful white-sided dolphins and Dall's porpoises.

On the fifth day, we headed our kayaks towards home. Retracing our route, we crossed the sound towards Orca Lab, a small orca research center, and headed back around Hansen Island. On the way we paddled past the "The Orcas' Lunchroom," a favorite haunt for seals and sea lions. As we rounded the top of the island, we hit our first real rain shower. Paddling hard, we headed into a small cove to put on our rain gear. Then, in a steady downpour, we recrossed Johnstone Strait and set up our final camp on a wide spit of land that jutted out into the channel. That evening, as we huddled under a wet tarp, more than one person reminisced on the fine qualities of a hot shower.

I knew we only had one more day of paddling. Would tomorrow be the day to see the orcas? I sent out a quick query. *Why don't you come to us?* is what I heard back. *We're having a huge party. The fishing is great!* Knowing that one of the orcas' favorite pastimes, as with dolphins, is catching and eating fish, I sensed they were celebrating and enjoying a lot of socializing. I wondered how I could get that far down Johnstone Strait. About fifteen minutes later, one of the guides struck up a conversation with me. "We were thinking," she said, "that you're not quite finished with the orcas." In a prior conversation I had talked to her about being able to identify the orcas by their dorsal fins. "Sometimes I can identify them," she had said. "I can usually tell who the B36's are. They're a group of three big males who usually travel together. They all have very tall straight fins. But why don't you go out on a whale boat tomorrow?" She continued, "We know just the one. The family who runs it has a long history with the orcas, and if anyone is going to find them, they will. It's a very fast boat."

Our last day was spent under cloudy skies. Finally, we turned and headed back to dry land. There was no sign of the orcas.

Early the next morning, I climbed aboard a trim whaleboat docked down at the harbor in Port McNeill. "We'll be heading for Victoria today," laughed the owner. "The orcas are a long ways off." I settled in for the ride as the boat sped down the strait. Apparently, the boat had been refitted so that the sound of the engine was reduced under water and the impact of the noise on the whales was lessened. When the boat slowed down, I climbed out onto the deck.

Everywhere I looked, I could see dorsal fins of orcas.

Furiously, I began snapping pictures. Then, just when I thought I'd seen the best, a shout would come from the other side of the boat, and everyone would run to the opposite railing. The pod was "I" pod, and one of the orcas, I41, had a distinctive curl at the top of his dorsal fin. The rules of whale-watching in these waters try to limit people from getting any closer than 100 meters, but every once in a while an orca would swim right underneath us or alongside, making it impossible for us to move away without disturbing them. It was an extraordinarly exciting and exhilarating afternoon. As we started home, I realized that I had in fact seen orcas on the last day. Now I *did* feel complete.

Pod of orcas in Johnstone Strait.

The allure and majesty of the orcas would stay with me. What would also remain was the experience of having seen so much other wildlife — dolphins, porpoises, seals, sea lions, otters, bears and eagles. Having been completely away from civilization for a week, I felt as though my senses had been expanded. I was listening in a new way. I was noticing subtle changes in the color of the water and cloud patterns. My eyes instinctively scanned the edges of the shoreline and the treetops. They roamed the horizon for unusual splashes or spouts. The world felt bigger. I felt alive.

CHAPTER 15

The Indigo Children

In many ways the Age of Light will be the Age of the Magical Child.... It is a realm of miracles and possibility. The image of the Dolphin swimming with a child will become a key symbol.... The child signifies...innocence, wonder and power.

— Celia Fenn, from *A Message for Lightworkers*

As my interest in the life of marine mammals grew, the most intimate interactions, both in and out of the water, continued to be with the bottlenose dolphins in the Caribbean. One evening on one of the dolphin encounter trips that I led with Jon, a group of us met in one of the marine biology institute's classrooms for storytelling and guided imagery processes. I described to the group my first "encounter" with dolphins — an overwhelmingly powerful experience that had occurred in nonphysical form while I was in meditation.*

I told them how the compelling force of that experience had led me to seek out dolphins in the wild and, finally, to having intimate contact with them in the tranquil setting of the Institute here on the island.

---

*This experience is fully described in the story "Close Encounter," Chapter 2.

The result of these encounters was that I had become deeply interested in the transformative effect dolphins seemed to have on humans. I read and listened to stories about people having startling and seemingly miraculous interactions with dolphins. I met people who believed they had received assistance or guidance from dolphins in an energetic or spiritual way. Others said they had been healed of serious illness, had awakened to new levels of insight, or had met a future romantic partner while swimming with dolphins.

These stories, with all their seemingly outlandish qualities, only served to heighten my interest and curiosity; *and* they left me with more questions than answers. How were people receiving these communications? Could dolphins "read" us beyond what they were able to "see" with their echolocation? Were some or all dolphins connected to some form of spiritual intelligence that enabled them to act as guides or teachers? Or did they represent a wiser, more compassionate part of ourselves that was simply mirroring itself in the outside world? I suspected that the answer to many of these questions was "yes." But just how these answers were revealed continued to vary and to surprise me.

Literally and figuratively I knew I was diving in deep water. Information that is received through psychic and intuitive awareness is difficult to test and prove, but often in life, my body and inner being have sensed and known the truth of experiences that the left side of my brain has not always been able to analyze or understand.

☙☙☙

As we came to the end of our evening sharing, Jon
and I suggested that we finish with a guided meditation.
We wanted people to slowly shift their attention away
from conversation and discussion to a place of inner
stillness and listening. I could feel the fatigue of some
people and the difficulty they were having focusing. I
noticed how the thoughts and conversations from the
day kept flooding through my own mind. Finally, a feeling
of expansiveness began to open up inside me. I took a deep
breath, and then another. I concentrated on breathing in
and breathing out. I started to sink down and settle within myself. "See if there is a particular dolphin that would
like to meet you," I heard Jon say. "Ask, if there is anything
you need to know." So far, there had been nothing of signif-
icance beyond the reruns of my own thoughts.

Then, suddenly, I saw a peculiar picture come into
focus inside my head. Instead of seeing one or two
dolphins, I saw a small school of about four miniature
dolphins swimming, not in the water, but vertically —
straight up my spine! I was taken aback. What were they
doing there?

I knew immediately that the dolphins I was "seeing"
were the younger ones from the island. Most of my time
at the resort had been spent with the youngsters of the
pod. They ranged from nine months to a little over two
and a half years old. They consistently exhibited an
openness and fearlessness that was deeply compelling —
and irresistible. Two of them, Maury and Li'l Bill, I had
met the first year we had come with a group. Now they
acted like young adults, participating fully in encounters
with people along the shore. Another skittish young
female, Mika, had blossomed into a calm and gentle
sweetheart. I attributed her change in attitude to the

careful handling of her young trainer. The other young male dolphin, Ronnie, was an up-and-coming star — flirtatious and friendly with everyone — much like *his* trainer.

Two of the dolphins that I had bonded with previously, Tela and Dexter, had been moved to other facilities. I grieved over the loss of Tela. If there had ever been a dolphin to fall head-over-heels in love with, it was her. But I had known that Tela and Dexter were among the most capable and adaptable of the pod, and therefore likely candidates for transfer. I wished them well in their new homes.

⌒⌒⌒

Some of my most memorable interactions with the younger dolphins occurred in the early morning, just after sunrise, when the sky glowed a soft pink, and the only sound was the gentle twittering of birds. I would slip on my flip flops and silently walk past the sleeping occupants in the other bungalows, and out to the taxi-boat at the end of the key. Then, instead of heading toward the restaurant for an early breakfast, I would ask the driver to take me out to the other tiny islet in the bay, where the dolphins were. The dolphins lived in a large enclosure surrounded by a wooden catwalk off this little island. Waving good-bye to the driver, I would get down on my stomach on the smooth gray planking and begin splashing the water with my fingers.

Invariably, within a minute or so, one or more of the youngsters would show up. Often they brought "gifts" or "toys" to play with — strands of sea grass, a leaf, or a small rock or shell — carrying them in their mouths or on a flipper. They soon had me well trained, and I would arrive with a handful of fresh grass or a large shiny yellow leaf. All kinds of games evolved from this morning ritual — "catch" and "fetch" and "keep-away" were a few. I remember one morning in particular. Maury showed up. Daintily she approached me carrying a beautiful piece of white coral as a present. What put me into hysterics was what she had on her head: a golden crown of tangled seaweed draped itself demurely down over her eyes. On a subsequent day Li'l Bill was the first dolphin to arrive. For no apparent reason, he immediately did a series of five full-body breaches in a circle right in front me; then, just as suddenly, he turned and disappeared.

Playing with Ronnie and a piece of seagrass off the catwalk.

A few days after my experience with Maury, I was in a dolphin-training session with her. There were six of us and a trainer at one of the platforms, and we were learning the hand signal for jumping. The right arm and hand hangs down and then swings upward across the body and is held straight above the head. As participants took their place, one by one, at the edge of the platform to practice the signal, Maury performed beautifully. When it was my turn, I swung my arm high into the air. Maury disappeared under water and then surfaced a short while later, carrying a small piece of coral in her mouth. I didn't know whether to be thrilled or mortified. "Do it again," said her trainer; "she didn't do it." Again, I did the hand signal — perfectly, I thought. Again, Maury vanished and then reappeared, tidily carrying a small gift. The trainer was firm. "Turn around and ignore her." I cringed as I turned my back on her. What I really wanted to do was get down on my knees and give her a big hug and thank her for remembering me and for bringing me her tiny presents. But this was a training session, and she was being asked to execute a specific command. So after ignoring her for almost a full minute I turned and, facing her, swung my arm into the air a third time. Maury took off through the water and performed a beautiful, perfect jump right out in the middle of the lagoon. I didn't know whether to laugh or cry, I was so proud of her.

On another occasion, at a dolphin encounter session along the shore, one of the senior trainers invited me over to watch as he worked with a tiny new dolphin named Fiona. Her sweet little face peered inquisitively up at mine as she allowed the trainer to briefly stroke her while she brushed past us. "She's just getting used to human touch," he explained. Fiona was nine months old. As I watched her diminutive form dive and resurface, I suddenly remembered having seen a dolphin playing by itself, tossing a stick through the water. As I had this thought, Fiona returned, carrying a stick! I reached and, easing it from her mouth, tossed it into the distance. She turned, splashed through the water, grabbed the stick and brought it back. "Would you mind babysitting for a few minutes?" said the trainer; "there's something I need to check on." I felt as if the doors of heaven had just been opened and I was being asked to take care of an angel. The next ten minutes were an indescribable dance of wondrous and intimate connection. With fierce rapidity Fiona learned how the simple tossing and retrieving of a stick was a magical way of making contact with another being. Her excitement and pleasure at having discovered a whole new world were a palpable experience of profound joy.

⌒⌒⌒

When I "asked" the young dolphins I had seen in my meditation what I needed to know, they advised me to *listen to the children.* Some of those being born on the planet now would be teachers of higher wisdom. Earlier in my life, I had worked in the Waldorf schools, founded by the German educator and visionary Rudolf Steiner. I had been inspired by their philosophy, which is based on a holistic picture of man, addressing the physical, emotional and spiritual aspects of our development. Now, I felt like I was being invited to revisit that period of my life.

I also found myself intrigued by some of the curious synchronicities that surrounded the "suggestions" I was receiving from the young dolphins. For example, just before leaving for our trip I had met the young son of a woman visiting from out of town. When the boy and his mom arrived at the house, he walked up to me through the front door, looked me in the eye, shook my hand and politely introduced himself. Then he went directly to my room where he saw the large stuffed dolphin on my bed, picked it up and asked if he could sleep with it that night. I was surprised by his directness, but impressed by his sense of self and knowing what he wanted.

In another instance at the resort, I noticed a family who seemed to be orbiting the periphery of our group, especially when we were interacting with the dolphins. The young parents were from Europe. Their son was so advanced for his age that they were unsure about what to do with him. The boy had shown a particular interest in our group and in the work with the dolphins. Jon suggested he join our group one evening. The father was reluctant to have his son join us, explaining that he'd probably embarrass us with the extent and complexity of his questions.

A younger dolphin "smiling."

These children, and the image I'd seen of the four young dolphins traveling up my spine, caused me to think of the Indigo children. The Indigos have been defined by Wendy Chapman, a teacher and program supervisor for gifted children, as a large percentage of the generation of children being born today.* They are characterized as often being rebellious, extremely emotional and physically sensitive, highly talented (academically or psychically), very empathic and compassionate, and wise beyond their years.

In the book *The Indigo Children,* by Lee Carroll and Jan Tober, they are described extensively. Indigo children are independent, and show great compassion for other living beings, including animals, plant life and other people. Eating is not such a big deal to them. They are sensitive to cruelty and unfairness, are usually generous, have high levels of intuition and are often counselors to their friends and family. Usually they know what they want and go after it. They have a strong sense of responsibility, adhering to guidelines rather than rules, and most of their decisions and interactions come from a place of love.

I had also come to know about these children through the work of James Twyman, author of *Emissary of Light: Adventures with the Secret Peacemakers* and *Emissaries of Love: The Psychic Children Speak to the World,* both of which I had read some years ago with great interest. A musician, Twyman was guided to write the music for some of the great "peace" prayers from the world's leading religions. Later he traveled to the remote and mountainous regions of war-torn Bosnia and Croatia where he played his new music and made contact with a mystical community called the Emissaries of Light. In 2001, he was led on a journey to Bulgaria

---

*Wendy Chapman is the Director of the Metagifted Education Resource Organization, www.metagifted.org, the goal of which is to educate people about giftedness — both academic and metaphysical. Metagifted is particularly focused on helping Indigo Children.

where he met four psychic children living in a mon-
astery. Their message to the adults of the world was
phrased in the question, "How would you act and what
would you do if you knew that you were an Emissary of
Love right now?"

That question brings to mind the first song on one
of Twyman's CD's, which is the well-known Prayer of
St. Francis. It begins: "Oh lord, make me an instrument
of your peace. Where there is hatred let me bring love."

According to Wendy Chapman, the name Indigo
refers to the deep-blue color that, in many spiritual
traditions, is associated with the third-eye chakra, the
center of psychic or clairvoyant activity in the human
aura. In many Indigos this faculty is believed to be wide-
open, allowing some penetration of the "veil" that
separates our world from the world of spirit.

I had always thought of the melon, or large forehead, on a dolphin as a kind of giant third eye. In echo-
locating, it appears that dolphins focus and beam their
echolocation clicks through this frontal fatty tissue and
then out in front of them. Sounds reflect and echo back,
giving the dolphin information about the size, shape and
density of objects. This gives the dolphins powerful imaging capabilities. My feeling is they are probably pro-
cessing a lot more information as well — not as linear
concepts like we often do, but in a more multidimen-
sional way, including, perhaps, "psychic" information
(feelings and knowingness coming through pictures).

I remember my first experience with dolphins in the
wild in Hawaii. My inner vision had been running a full-
color movie the whole time I was there. It was clear to
me that this inner picture show was influenced somehow

by the dolphins' proximity. The "movie" seemed to override many of my usual thought patterns.

⌒⌒⌒

In conclusion, perhaps my "message" from the young dolphins — to *listen to the children* — is not only to attend to the voices of the younger souls on this planet, but also points to the need to awaken the wise child within all of us, for the sake of the planet as well as ourselves. We could certainly use more love and compassion in this world. In fact, our very lives and our future depend on it.

When I picture the four small dolphins traveling up my spine, I am reminded of the raising of the kundalini, or spiritual energy, to the crown chakra; of the desire of all true seekers for enlightenment; of the power of love expanding through community; and, finally, of the journey with others and with the dolphins into a more light-filled and compassionate way of existence.

Maury.

CHAPTER 16

Mother's Trip to Paradise

Love doesn't make the world go 'round.
Love is what makes the ride worthwhile.
— Franklin P. Jones

One Sunday morning, not long after Christmas, I received a telephone call from my mother in Norway. She wanted to come to the States for a visit. She had returned to her native homeland with my father, and although they would later separate, she had continued living there. Her request to see me brought up a myriad of inner responses, even though I am a grown woman living my own life and I live on the other side of the globe. Perhaps these stem from my early childhood, when I longed to see more of my father who was often absent, and instead felt and reacted to the predominant presence of my mother. With the passage of time, the relationship between my mother and me has deepened — I have come to enjoy our long conversations over the phone, and on occasion I even ask her advice on some problem I'm mulling over. But the immediate questions were, how long would she stay, and would it be with me or my sister?

My sister's position was clear: It was not realistic for my mother to stay for more than a week with her. With my sister's work schedule, even that was stretching it. Then she had an idea. Mother could come on the dolphin trip. I knew immediately that I'd been one-upped. Even at the grand old age of almost eighty, my mother still loved to travel. In fact, she had always been a hard act to follow. After her divorce from my father, I would periodically receive her postcards — they came from such far-away and exotic places as Kenya, Peru and Bhutan. On one of her trips a few years ago on a pilgrimage to the city of Santiago in northern Spain, she had met a Norwegian man and fallen in love. They have become close companions, sharing friends, and traveling when time permits. For this trip he had stayed at home, not one to enjoy the tropics.

As I started researching flights from Norway to the Caribbean, I wondered how I would manage a large group of participants *and* my mother. As it turned out, my anxiety was unwarranted. Mother settled into a bright and spacious bungalow a few doors down from me. While I was off communing with the dolphins in the early hours just around sunrise, she was taking her morning swim. She would arrive at breakfast showered and neatly dressed, her hair coiffed and styled, with her snorkeling gear in a colorful beach bag slung over her shoulder. At mealtimes she kept up a constant stream of conversation, introducing herself proudly as Mom.

Towards the end of our week the group participants

had graduated to learning how to take rides from the dolphins. A dorsal ride consists of holding onto the dorsal fin with both hands and then allowing the dolphin to pull you through the water. There were shrieks of laughter as everyone, including my mother, took turns. As she made a wide circle while holding onto the dolphin, she gallantly let go with one hand and waved at the photographer.

My mother holds the dorsal fin of a dolphin for a ride.

Another ride is the toe push. A person floats facedown in the water with the soles of their feet pushed away, as if pressing up against a wall. The dolphin places its rostrum against one foot and then propels the person forward, allowing the upper body to be lifted up from the water with both arms stretched out to either side. When my mother's turn came, she lay face-down in the water with her feet stretched behind her. Moments after the dolphin made contact, she was tearing across the bay at high speed. We cheered her on. Soon, however, it became apparent that she was not making the nice half-circle back to shore like everyone else. "Turn, turn," we yelled. That was the one piece of instruction she had missed. The dolphin would push, but it was up to the person to steer by leaning in the direction they wanted to go. By this time my mother was more than halfway across the lagoon. There were audible groans from people envying the length of her ride. Finally, my mother and the dolphin came to a crashing halt way out in the middle. Somehow my mother managed to turn around, line herself up and have the dolphin push her back to shore in a great cloud of water. We cheered and clapped enthusiastically.

*One of the participants, Sabine, with my mother
and me (from left).*

A few days later, at a closing evening ceremony under an endless sea of blinking stars and a huge yellow moon, and surrounded by tiki torches, each participant was given a chance to share a few final words with the rest of the group. This had always been a special and heartfelt time of sharing. When it was my mother's turn, she moved to the front, graciously thanked everyone for their participation and said she had enjoyed herself immensely. She commented humorously on her mad ride across the lagoon, reminding everyone again that she was almost eighty years old. "When I think of the long ride I've had through life," she concluded, "I've come to realize that some of the most important things to remember are enjoying the warm sun above, the water below and the companion at my side. Thank you for a wonderful week." She was given a stirring round of applause, having touched the hearts of all who'd spent time with her. I felt moved and very proud.

CHAPTER 17

Gentle Giants

I'm not a very sentimental person. I don't think we ought to save whales because they're cuddly or pettable. But it's simply an amazing experience having those whales roll over and look at you eye to eye. There's really an interspecies contact there. There's an intelligence.... That's about as far as I want to go with that, but it's... extraordinary.

—Robert F. Kennedy, Jr.,
National Resources Defense Council (NRDC) Lawyer
(from *Eye of the Whale* by Dick Russell)

I watched as a long, thin, gray-knuckled ridge rose to the top of the water from the aquamarine depths below. Like some great primordial water dragon, it slowly widened to reveal a patchy gray and white exterior that extended about eight feet across and the length of two boats. It was a California gray whale that had just surfaced about nine meters to the port side of our *panga*, a small motorized skiff operated by one of the locals.

I was in Magdalena Bay, near Boca de Soledad on the Baja Peninsula. The bay is one of three lagoons on the Pacific side that have been safeguarded as sanctuaries for the gray whales who migrate here every year to breed and birth their young. It was February, and I was here on a weeklong expedition to kayak along the mangrove estuaries and to see the whales.

In the summer of 2004 I had traveled up to the lush isles of British Columbia on a paddling adventure with the orcas. Now, six months later, opportunity had brought me south to the cacti and arid vistas of Mexico. I had heard about the "friendly" whales of Baja — the mother whales and their babies, especially — who would approach visitors in their boats seeking contact. Here was my chance to experience firsthand the truth to these whale tales. I was curious to see how the whales' interest in humans compared to that of dolphins.

There is a remarkable story about how a human is first reported to have touched one of Baja's "friendly" whales. In 1972, a man named Pachico Mayoral and a friend were out in a small fishing boat in Laguna San Ignacio, another wide bay along Baja's western coast. Quite unexpectedly, they were approached by a large gray whale that rubbed up against the side of their boat. At first, the two men feared for their lives, but as the whale continued its attentions, Pachico finally reached over the side and touched it. Later, he talked of this event as being a profound and life-changing experience, similar to that of holding his first-born child. The gigantic whale seemed to relish the contact and continued to submerge itself and reappear on the other side of the *panga* for at least another two hours. Pachico returned to his village, and the story spread about how the two men had touched a whale. This was the beginning of a whole new change in attitude toward the whales and the start of a new industry in whale-watching and whale-petting.

☞☞☞

The story of the first "friendlies" stayed in the back of my mind as our group of eleven kayakers and three guides quietly paddled by the tangled roots of red and white mangroves. We were headed for our campsite on a remote and windswept island in Bahia Magdalena. Great white heron and snowy egrets roosted in the thick branches on either side. Pelicans, ospreys, cormorants and gulls flew overhead and lined the edges of sandy coves. In the distance we could see the feathery spouts of whales as they surfaced for air. As we skirted a final bend in the channel, we saw the rounded white domes of two spacious tents on a wide beach. We had arrived at what would be our new home for the next week.

Tomorrow would be our first day on the water for a closer look at these gentle giants. Instead of using our kayaks, we would be adhering to whale-watching regulations by traveling in two open-air *pangas* driven by licensed operators. That evening, as we sat in a circle under the flickering light of a lantern in one of the domes, we heard a retelling of the account of the first "friendly" whale. As the story ended, I knew we were in for a rare treat when we learned that our guide Poncho was one of Pachico's sons. Growing up as a fisherman, he had witnessed the gradual shifts that had taken place in his community as a result of the changes in attitude towards the whales and as awareness grew for the need to preserve the environment. This led him to work with several organizations whose aims were to protect declining species and to train locals in language and leadership skills. Today, he does what he likes best — being a guide and introducing people to the beloved *ballena gris,* or gray whale.

As the evening continued, and over the days ahead, we familiarized ourselves with whale facts and history. For centuries, the gray whales were hunted — first by the native hunters of North America and Asia, and later by the Yankee and European whalers. Whale bone and baleen were used to make hoop skirts and corsets, and although the oil from gray whales was of poorer quality than that of other whales, it increased in value as the whale population declined. With the evolution of whaling technology, and the use of explosive harpoons and steam-powered ships, it was estimated that less than 1,000 Pacific gray whales were left by the 1930s. With extinction imminent, an international agreement was finally reached in 1946 that banned all commercial whaling. The treaty was signed by most whaling nations. Today, the Pacific gray whales have undergone an unprecedented recovery, with an average annual increase of 2.5 percent.

Gray whales are part of the subclassification of cetaceans known as the Mysticetes, or baleen whales (which use a fine rubbery fringe on the upper and lower jaws to filter plankton and other tiny aquatic animals). This is different from dolphins, who are classified as Odontocetes, or toothed "whales." Toothed whales are considered predatory, feeding on fish or other marine mammals. For the most part, the gray whales graze in shallow waters, scooping up mud along the bottom, which is then pushed and filtered through their baleen with a giant tongue. Small invertebrates are left trapped inside their mouth and then swallowed.

Another peculiarity of the grays is the fact that they are hosts to several species of barnacles and amphipods, or whale lice, which feed on the whale's skin. These tiny crustaceans give the gray whales their characteristic mottled appearance. Born a shiny black, the whales are soon discolored by large distinctive patches of crusty white. The small parasitic creatures that attach themselves seem to assist the whales by keeping them clean.

An additional characteristic of the baleens is the presence of two blowholes on the top of their head, as compared to the one blowhole of a dolphin. When a gray whale surfaces to breathe, it exhales with a tall heart-shaped spout. Appearing like a faint mist from afar, this watery plume actually rises up to over twice the height of a human. Later that week, some of us would be lucky enough to see the rainbows that sometimes blink in the sparkling haze of the whales' breath.

Finally, the gray whales do not have a dorsal fin like dolphins. Instead, a series of six to twelve bumps traces the length of the whale's spine, making them at first glance look like some ancient and primitive creature that has come back to life from another age.

What the Pacific grays are perhaps most known for is their extraordinary migratory route, one of the longest of any species of mammal. Taken each winter from the nutrient-rich arctic waters of the Chukchi and Bering Seas off Alaska, they traverse the entire length of the North American continent to the southern reaches of the Baja peninsula. It's a round-trip journey of about 10,000 miles — almost two months of travel each way. To prepare for the trek, the whales accumulate from six to twelve inches of extra blubber over a summer of intensive feeding. Then, with the gestation period of a baby whale being about twelve months, the goal of the mothers by midwinter is to reach the tropical waters of the south where their young can be safely born.

The author petting Bubbles.

≈≈≈

On our first morning of whale-watching, we gathered in the faint early light wondering what lay ahead. After pulling on our life jackets, and loading up on our cameras and film, we motored out into the center of the bay. There, among a euphony of spouts, we had our first close-up view of the long, speckled bodies of several mamas as they rose to the surface, and nearby their babies, whose small heads looked like little antique dinosaurs with long downturned mouths etched along their jaws. Gingerly, we puttered closer, transfixed by the gentle rise and fall of their giant bodies and the soft whooshing sound of their breathing. One mother/baby pair, who particularly caught our attention, quickly received the nicknames "Scarface" and "Bubbles." The mama had a distinctive white scar near her blowhole, and her baby, we noticed, seemed to enjoy blowing big,

Poncho stroking the baleen of the whale.

round bubbles that came tumbling up to the surface in short bursts. A few people were able to just reach the whales from the boat for a brief touch. Otherwise, the whales remained a discrete distance away from us.

As our session went on, we navigated out to the choppier waters at the bay's mouth, where a number of juveniles frolicked in the waves, breaching and spy-hopping. I couldn't help wonder, as I had with the dolphins and the orcas, how much their behavior was purely for themselves and how much might have been for us. Either way, the air was soon filled with our delighted shrieks as these giant leviathans came crashing down in thunderous displays of water theatrics, or as they stood vertically, apparently eyeballing us with curious consideration. Some of the whales would swim alongside us about ten meters out before disappearing into the watery depths. On our return trip we cruised into what appeared to be a whale nursery, a quieter area off one of the islands. Several mothers floated motionless for long periods of time, as if they were sunbathing. Then, we would see the small spout of a baby next to one of the mamas, and we guessed that her calf had been nursing.

On the day of our second excursion, there was a heightened feeling of anticipation among the group's participants. Traveling with Poncho, our boat soon made contact with a mother and baby. Interestingly, the mom did not interact, but hovered protectively nearby as her youngster bumped up against the *panga* and went from one person to the next receiving scratching and caresses. Its rostrum was covered with circular white barnacles and short little hair bristles. When it was Poncho's turn, he ran his hands along the baby's mouth, which then opened to reveal a feathery fringe of baleen that he combed with his fingers, much to everyone's delight and amazement. Soon afterwards, we encountered Scarface and Bubbles again. While Bubbles interacted with each of us along the side, his mom swam slowly back and forth beneath our boat, rubbing her back against the hull.

Later, as I rinsed my hands in the surf after a picnic lunch, a sudden plopping sound caused me look up and see the graceful silhouettes of two dolphins leaping out in front of me. It was as if they were sending a quick "hello," wishing me well with their large mammalian friends.

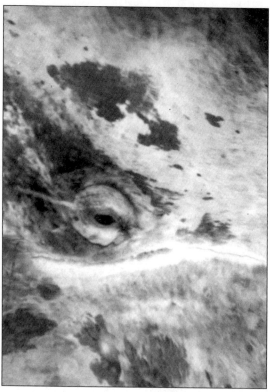

Eye of the whale.

On the last day of our whale adventures, a few of us decided to pay our driver for some extra time. We headed back out into the lagoon with Poncho, knowing that this would be our last encounter. Almost right away, we met Scarface and Bubbles. With great enthusiasm, Bubbles bobbed from person to person. As he slid along the edge of the boat, sideways in the water, I could see his eye looking directly into mine. Despite his size, I felt profoundly struck by his vulnerability and apparent trust. Shyly, he opened his mouth and allowed me to stroke his tongue. I had the feeling that if given the opportunity, he could easily learn to recognize simple gestures and respond to certain voice intonations, much like bottlenose dolphins.

As our motor idled, we were joined by several other *pangas* seeking a closer look. Perhaps most impressive, the whales took care to greet each of the newcomers, swimming from one boat to the next, giving everyone a chance to pet or stroke them. Several times, as the mama approached, her giant spout would douse the passengers with a cloud of whale's breath as they shrieked with laughter. A couple of the *pangas* carried Mexican visitors, and it was particularly endearing to see them, with ear-to-ear smiles, reaching over to pet what had once been known as a fearsome devil-fish.

As we headed back to camp and I stared out at the large expanse of bay with the many spouts that appeared at regular intervals, I couldn't help but wonder how it was that we had managed to find the same whales on each of our excursions. Did the whales recognize and remember us from one day to the next? Did they seek us out especially? Of all the whales in this vast lagoon, were there only a few interested in making contact with people? Days later, when I perused a wall full of colorful whale photos for sale, I found myself saying, "No, that's not Bubbles. That's not him." I left the kiosk empty-handed, relying instead on the photography skills of myself and our group.

⌒⌒⌒

At the end of the week, we loaded our gear into our kayaks and later traveled by van through the desert land-scape of central Baja to the city of Loreto, a small tourist town on the Sea of Cortez. Gradually, we re-entered a more "civilized" way of life, with our metal vehicles and glass window panes, our currency and our ice cream. I fit myself back into the closed security of four walls, hot running water and flush toilets. As much as we humans have come to enjoy our creature comforts, I knew that a price comes with it — the risk of forgetting our intimate connection to the rest of the natural world.

As I boarded the plane, I took in a few more long breaths of warm, moist air. Then, gluing my face to the tiny window by my seat, I looked for the delicate spouts of the whales in the sparkling sea below me. Inwardly, I thanked them for offering me an experience of such dis-arming trust and vulnerability. If the whales harbored any residual memories in their cells or DNA of their earlier relationship to man, they seem to have decided to forgive their former enemies, offering instead a new way of relating that was both compelling and profound. I decid-ed to keep my personal snapshot of Bubbles safely tucked away in my heart.

CHAPTER 18

Dancing on Water

> *Meditation means participation in the celebration of existence. Do not just be a spectator; participate in the mystery of life. Dance it, sing it, feel it, be it.*
>
> —Osho, spiritual teacher

An area of life that remains mysteriously challenging to many of us is our intimate relationships. As human beings, we appear to be uniquely individual and separate from one another, and yet inherent in that individuality is a longing for connection to the other. Somehow, we intuit a greater wholeness that includes deep intimacy and seeks to unite many of our differences. By their profound example, our cetacean friends appear to offer us significant clues and hope in our quest for Oneness.

On one of our trips to the Caribbean, a small group of us had watched a spontaneous showing of the IMAX movie *Dolphins* on DVD at the institute. The cinematography in the movie is stunning. There are beautiful shots of dolphins from all over the world. Some of the most impressive footage is of the dolphins moving together — turning, spinning and flowing — dancing in perfect harmony. Since the world of dance in its many forms is particularly close to my heart, I felt especially moved by the simple and wonderful chemistry of the dolphins' movements. Taking their cues from one another, they moved, seemingly without effort, in and out of beautiful and unrehearsed relational patterns — blending, mirroring, imitating — their individual bodies instruments of a larger and much more complex composition. With exquisite grace and sensitivity, they became a kaleidescopic sea of changing shapes and forms that constantly shifted and rearranged themselves. The simple beauty of it was breathtaking. If only we as humans could move as fluidly, staying with our knowingness and our body's intelligence, listening and attuning ourselves, adapting with grace and ease to life's constant changes.

I feel as though the dolphins continue, by way of example, to invite us to dance — to let go and to keep moving. So often it seems as though we create our own obstacles and resistances in life out of our fear of change. "If I don't work to hold on, I'll lose this person or that part of my life," we think; whereas, if we can learn to trust ourselves more and pay attention to our deepest truths and longings, and let go and actually breathe, the river of life can keep moving forward. Otherwise, we risk getting stuck and can feel blocked in the expression of our feelings and in the choices we make or don't make for ourselves. We need to feel revitalized by new oxygen and energy.

Often, by letting go of the need to control or force a situation, I've learned that we allow things and people to come towards us willingly and in their own time. We create breath and space around ourselves energetically where something new can be born. For me, sometimes this has meant spending time alone. It has been about staying with myself and discovering my own joy. The dolphins, especially, remind us that all of us as living creatures are intrinsically rooted in a joy just waiting to be tapped. In our essence, we are love and joy. Ultimately, it's not something outside of ourselves that needs to be discovered and found — something that we then try to own and contain. It's something that we already have, and are.

Another teaching for me that seems to be expressed by both the dolphins and the whales, and that speaks directly to our longing for relationship, is that of staying in present time and engaging in what appears to be the practice of forgiveness. The inability to forgive, it seems to me, is often at the core of humanity's greatest suffering. My intimate encounters with the dolphins and whales have always been a source of great joy. Time and time again they show up, seemingly without judgment, and entirely accepting of who I am as a human. Despite the terrible atrocities that we as a race have inflicted upon their species over the millennia, they continue to seek us out as friends — helping, inspiring and entertaining us. If only we could rise to that level of appreciation for each other as human beings, perhaps the world would be a very different place.

In my own life, when Love has shown up and invited me as a guest to its table, I have also, paradoxically, been faced with my own greatest fears and inner challenges. As love is revealed, so have I felt the shadow emotions of jealousy, fear and anger. The allurement of love and joy asks us to stay in our heart and to take up the practice of forgiveness. I've had to learn that when I hold on to anger or hurt, to being "right," often I am being most hurtful to myself. There is great power in choosing to stay in our vulnerability, in willingly stepping into the unknown where nothing is certain except the surrendering to Love.

Near the end of one of our dolphin trips, during one of our group sharings, a woman described a beautiful insight she'd had. She said that every time she saw one of us touching a dolphin she felt as if she were witnessing a holy event. Surely, these are moments of heightened joy and awareness. I am reminded of the meaning of the Sanskrit word *satsang,* which means to associate with the divine or to be in loving communion with a Master. Simply being in the presence of a dolphin or whale is like receiving a transmission. It's as if they are functioning as a catalytic agent to remind us of who we really are.

"Laughing."

In its truest form, intimate relationship, like being with the dolphins, is similar to a spiritual practice. It is about attending to the moment at hand and committing to love. Instead of revisiting the past or racing forward into a still-to-be-revealed future — frightened of losing oneself or of being abandoned — we can learn to embrace the changing moods and seasons of our own inner and outer worlds. By choosing to stay in the dance, to ride the waves when they come, and then rest in the stillness, we create a sure place for joy. I continue to tumble and play across the wave tops. I am learning to dance on water.

With humility and great joy,

Namaste

Poems

Tela

dolphin-child
sea angel
greeting me
  laughing
flipping fish high
  silver flying
  catching
splash —
  see me!

nodding
  yes, yes, yes
her fluke slaps the water
  play
  new game
giggling
  spinning circles round
silly
  whistling
singing
  dolphin-child

Bodies of Light

Two bodies of light rise up from the sea,
  exuberant,
  joyous,
Spraying,
  splashing water droplets
  wildly,
Flashing sunlight in every direction.
Then streaking home,
Gracefully stopping a hair's breadth from my skin,
  resting, breathing,
Turning belly-up,
  quietly receiving touch,
Suspended between palm and water.

Wise liquid eyes
  see me.
Who are you?
Sweet beauty
  dolphin

Day of the Dolphins

The dolphins came —
  the mothers and their babies,
  the males swimming in pairs,
Beautifully synchronized,
  perfectly orchestrated
  and unrehearsed,
Making a ring around the humans,
And they began to jump
  straight and high,
  and into the sun,
As they've leapt for eons,
Even before humankind
  arrived on the planet,
And Before and Now and When,
  become One time and One place.

The girl on the beach
  knows and rejoices
  in life's celebration —
Explosions of light,
  dance of the heart,
The waves
  and the wind,
  and the dolphins

Contact

The wizened mammal heaves her barnacled brow
   up from the emerald depths,
     as foaming falls of water cascade off her back.
Her mighty fluke moves beneath the surface,
   and she propels her giant offspring forward.

Her baby rolls from her side,
   shiny and black,
Blowing a ring of perfect bubbles
   that float across the waves.

With curious intent,
   Mama Whale eyeballs the humans in their tiny craft.
Patiently, and with a gentleness that belies
   her great strength,
She settles and waits
   as she's waited for cycles and cycles of time,
Her long mottled body
   stretching out across the water.

Her offering is her newborn calf —
   a gesture of peace;
Let bygones be bygones, she says.

She waits for a New Beginning,
   for contact —
She waits for the first human touch.

Eye of the Whale

The eye of the whale,
  wise, venerable, and aged,
  beckons

A well of compassion,
Holding, nurturing, and
  cradling our small spirits

Offering a brief respite
  from the ways of the world.

A silver portal
  of solace and light
  shining

A pure place of power
  where the ancient memories
  converge and congeal

And the future
  is made manifest.

Cycle of Life

And the humans began to wake up
  like little lights,
  fireflies
  blinking all around the planet,
And they remembered:
  We are not alone,
  We are connected to every other living thing.
There *is* only Living and Life,
  transforming
  recycling,
  constantly rebirthing,
Breathing in
  and breathing out.
When we disown any part of that Life,
  chaos ensues,
And the balance is upset.

So people remembered
  and began to reclaim stewardship
  of their home.
Some cared for the trees
  and the flowers,
Some the snakes
  and the lizards,
Others the deer
  and the wolves,
Still others the whales
  and the dolphins.

Once again children
   grew up playing
   and laughing.

Life freely given
   leaves no room for fear,
Death becomes an old skin
   shed and reborn
   in the springtime.
Once again the cycle was seen
   as a circle,
Whole and unbroken.

Rain of Light

Cold stars pour down
 a dizzying delight
 of luminous electricity

Drenching flesh
 and bone,
Ecstatically pulsing,
 igniting the blood

A floodgate breaking
 into the dark
 womb of night.

Evening Splendor

The gentle waves carve
  their patterns in the sand,
Exquisite etchings,
  finely chiseled crevices
  and feathered canals,
Adorned with gifts fom the sea.
This faint and fabulous countryside
  is slowly revealed
  as the tide recedes —
Scattered pieces of colored glass
  rubbed smooth,
Deep blue and bleached bits of shell,
Tangled clumps of messy seaweed
  smelling of old fish and salt —
All these treasures
  moving restlessly
As a thin veil
  of water ripples back and forth,
Catching the last light of day;
A solemn, simple pageantry
  marked by single standing gulls,
And the unhurried footprints
  of passersby.

Dolphn Encounters:
Adventure Trips to the Caribbean

For more information on week-long trips with
the bottlenose dolphins, please see
www.dolphinpress.com
or contact Karin at karin@dolphinpress.com.
Also see Jon's Web site, www.thefirewithin.com.